Fundamental Theories of Physics

Volume 210

The international monograph series "Fundamental Theories of Physics" aims to stretch the boundaries of mainstream physics by clarifying and developing the theoretical and conceptual framework of physics and by applying it to a wide range of interdisciplinary scientific fields. Original contributions in well-established fields such as Quantum Physics, Relativity Theory, Cosmology, Quantum Field Theory, Statistical Mechanics and Nonlinear Dynamics are welcome. The series also provides a forum for non-conventional approaches to these fields. Publications should present new and promising ideas, with prospects for their further development, and carefully show how they connect to conventional views of the topic. Although the aim of this series is to go beyond established mainstream physics, a high profile and open-minded Editorial Board will evaluate all contributions carefully to ensure a high scientific standard.

Yaakov Friedman · Tzvi Scarr

A Novel Approach to Relativistic Dynamics

Integrating Gravity, Electromagnetism and Optics

Springer

Yaakov Friedman
Extended Relativity Research Center
Jerusalem College of Technology
Jerusalem, Israel

Tzvi Scarr
Department of Mathematics
Jerusalem College of Technology
Jerusalem, Israel

ISSN 0168-1222 ISSN 2365-6425 (electronic)
Fundamental Theories of Physics
ISBN 978-3-031-25213-6 ISBN 978-3-031-25214-3 (eBook)
https://doi.org/10.1007/978-3-031-25214-3

This Springer imprint is published by the registered company Springer Nature Switzerland AG
The registered company address is: Gewerbestrasse 11, 6330 Cham, Switzerland

Preface

This self-contained monograph provides a mathematically simple and physically meaningful model which unifies gravity, electromagnetism, optics, and even some quantum behavior. The simplicity of the model is achieved by working in the frame of an inertial observer instead of in curved spacetime.

Our approach to dynamics is geometric, and by plotting the *action function* of a spacetime, one is treated to a visualization of the geometry. Using these visualizations, one may readily compare the geometries of different types of fields. Moreover, a new understanding of the *energy-momentum* of a field emerges.

The reader will learn how to compute the precession of planets, the deflection of light, and the Shapiro time delay. Also covered is the relativistic motion of binary stars, including the generation of gravitational waves and a relativistic description of spin.

The mathematics is accessible to students after standard courses in multivariable calculus and linear algebra. For those unfamiliar with tensors and the calculus of variations, these topics are developed rigorously in the opening chapters. The unifying model presented here should prove useful to upper undergraduate and graduate students, as well as to seasoned researchers.

Jerusalem, Israel

Yaakov Friedman
Tzvi Scarr

Acknowledgments We would like to thank Larry Horwitz, Rainer Weiss, Bahram Mashhoon, Dan Scarr, Tepper Gill, Tzvi Weinberger, Yakov Itin, David Hai Gootvilig, Elazar Levzion, and Chanoch Cohen for their comments.

Contents

List of Figures

List of Tables

Chapter 1
Introduction

This monograph on relativistic dynamics is guided by three major themes - *geometry*, *unification* and *simplicity*. Our approach is geometric so that our mathematics has physical meaning and content. We use the full strength of Galileo Galilei's Principle of Relativity (Fig. 1.1) with respect to the Lorentz transformations. Our dynamics is also unifying-gravity, electromagnetism, and optics are integrated into the same model. Our dynamics is as simple as possible and yet can explain all known relativistic phenomena.

1.1 Physics via Geometry–A Historical Perspective

The first person to think that the laws of physics should define the geometry of space was Bernhard Riemann (1826–1866) [70] (Fig. 1.2). Although best known as a mathematician, Riemann became interested in physics in his early twenties. His lifelong dream was to develop the mathematics to unify the laws of electricity, magnetism, light and gravitation. At an 1894 conference in Vienna, the mathematician Felix Klein said:

> I must mention, first of all, that Riemann devoted much time and thought to physical considerations. Grown up under the tradition which is represented by the combinations of the names of Gauss and Wilhelm Weber, influenced on the other hand by Herbart's philosophy, he endeavored again and again to find a general mathematical formulation for the laws underlying all natural phenomena .... The point to which I wish to call your attention is that these physical views are the mainspring of Riemann's purely mathematical investigations [65].

Riemann's approach to physics was geometric. In fact, he replaced straight lines with geodesics, an idea later used in General Relativity (GR). As pointed out in [82], "one of the main features of the local geometry conceived by Riemann is that it is

Y. Friedman and T. Scarr, *A Novel Approach to Relativistic Dynamics*,
Fundamental Theories of Physics 210,
https://doi.org/10.1007/978-3-031-25214-3_1

Fig. 1.1 Galileo Galilei (1636 portrait by Justus Sustermans)

Fig. 1.2 Bernhard Riemann

well suited to the study of gravity and more general fields in physics." Moreover, Riemann believed that the forces at play in a system determine the geometry of the system. For Riemann, *force equals geometry*.

The application of Riemann's mathematics to gravity would have to wait for two new ideas. While Riemann considered how forces affect *space*, physics must be carried out in *spacetime*. One must consider trajectories in spacetime, not in space. For example, in flat spacetime, an object moves with constant velocity if and only if his trajectory in *spacetime* is a straight line. On the other hand, knowing that an object moves along a straight line in *space* tells one nothing about whether the object is accelerating. As Minkowski said, "Henceforth, space by itself, and time by itself, are doomed to fade away into mere shadows, and only a kind of union of the two will preserve an independent reality [77]." This led to the second idea. Riemann worked only with positive definite metrics, whereas the Minkowski metric of flat spacetime is not positive definite. The relaxing of the requirement of positive-definiteness to non-degeneracy led to the development of *pseudo-Riemannian geometry*.

In 1915, fifty years after Riemann's death, Einstein (Fig. 1.3) used pseudo-Riemannian geometry as the cornerstone of GR. Acknowledging his reliance on Riemann, Einstein said:

Fig. 1.3 Albert Einstein (1921 photograph by F Schmutzer)

> But the physicists were still far removed from such a way of thinking; space was still, for them, a rigid, homogeneous something, incapable of changing or assuming various states. Only the genius of Riemann, solitary and uncomprehended, had already won its way to a new conception of space, in which space was deprived of its rigidity, and the possibility of its partaking in physical events was recognized. This intellectual achievement commands our admiration all the more for having preceded Faraday's and Maxwell's field theory of electricity [28].

GR is a direct application of "force equals geometry." In GR, the gravitational force curves spacetime. Since, by the Equivalence Principle, the acceleration of an object in a gravitational field is independent of its mass, curved spacetime can be considered a *stage* on which objects move. In other words, the geometry is the same for all objects.

However, GR represents only a partial fulfillment of Riemann's program. Since the Equivalence Principle holds only for gravitation, GR singles out the gravitational force from other forces which are object dependent. Take the electromagnetic force, for example. The acceleration of a charged particle in an electromagnetic field depends on its charge-to-mass ratio, implying that the electromagnetic field does *not* create a common stage on which all particles move. Electromagnetism is an object-dependent force. Indeed, a neutral particle does not feel any electromagnetic force at all. Thus, the way in which spacetime is influenced by an electromagnetic field depends on *both* the sources of the field *and* a single, intrinsic property of the object–its charge-to-mass ratio. This was also recognized in the geometric approach of [22]. This raises the question: Can Riemann's principle of "force equals geometry" be applied to other forces? Can Riemann's program be extended to object-dependent forces?

In this book, we introduce several new ideas which enable us to geometrize not only gravity, but also electromagnetism, motion in media, and even some quantum behavior. The model presented here is thus a continuation of Riemann's program.

For us, *geometry* means the measurement of distances between two infinitesimally close points in spacetime. This lies at the core of the theory because we use a variational principle to find *stationary* or "shortest" paths of objects through spacetime.

> Many results in both classical and quantum physics can be expressed as variational principles, and it is often when expressed in this form that their physical meaning is most clearly understood. Moreover, once a physical phenomenon has been written as a variational principle, ... it is usually possible to identify conserved quantities, or symmetries of the system of interest, that otherwise might be found only with considerable effort [86].

In pseudo-Riemannian geometry, the distance between two infinitesimally close spacetime points is quadratic in the temporal and spatial separations. This is a seemingly reasonable assumption and strongly supported by the Pythagorean Theorem and its generalizations. However, for the electromagnetic field, one requires a *linear* dependence on displacements. Moreover, the exclusively quadratic dependence was needed only to ensure positive definiteness of the metric. Once we no longer require this, one should allow the simpler, linear dependence. Thus, our first new idea is to *include both linear and quadratic dependence* in our *action function* – the function for computing distances in spacetime.

Second, by using geometries which are allowed to depend on the charge-to-mass ratio of the moving object, we can incorporate electromagnetism into our model. We can also model the motion of light in isotropic media and between different media. In these cases, the geometry depends on the photon's energy or frequency. Thus, the second new idea is *to broaden the notion of geometry to include object dependence*.

1.2 Unification and Simplicity

The second theme of this book is unification. A good theory unifies concepts that had previously been treated as distinct. Einstein attempted to unify gravity with electromagnetism, and his dream was to unify relativity with quantum mechanics. To be sure, we have not achieved full unification here. Nevertheless, we do have a unified approach to gravity, electromagnetism, optics and even some quantum effects. Thus, this book may provide a solid foundation for a complete, unified theory.

Our third theme is *simplicity*:

> Explanations that posit fewer entities, or fewer kinds of entities, are to be preferred to explanations that posit more. - William of Occam

> Scientists must use the simplest means of arriving at their results and exclude everything not perceived by the senses. - Ernst Mach

> A physical theory should be as simple as possible, but not simpler. - Albert Einstein

> If you can't explain it simply, you don't understand it well enough. - Albert Einstein

In physics, if two theories predict and account for the same phenomena, the simpler theory is to be preferred. On the other hand, a theory which accounts for

Fig. 1.4 Isaac Newton (1689 portrait by Godfrey Kneller)

more phenomena must be favored, even if it is more complicated. For example, Special Relativity (SR) replaced Newtonian mechanics (Fig. 1.4) because it explained Maxwell's equations, the dragging of light in moving water, and the invariance of the speed of light, while containing Newtonian mechanics as a limit. Ten years later, GR explained what SR couldn't-the gravitational redshift and the anomalous precession of Mercury's orbit. GR went on to explain the periastron advance of a binary star, the deflection of light, the Shapiro time delay and gravitational waves. For more than a century, GR has been the simplest, in fact, the only theory that explains these relativistic phenomena. Hence, any new theory must also explain these phenomena. It must, as we say, pass the *tests of* GR.

In light of the above statements of William of Occam, Ernst Mach and Einstein himself, it is natural to ask whether there exists a simpler theory which also passes the tests of GR. After all, GR is a complicated theory. Working on a curved spacetime manifold, one is immediately greeted by covariant derivatives, Christoffel symbols, curvature tensors, and, last but certainly not least, the field equations-a system of non-linear, partial differential equations with *ten degrees of freedom*. This is not an easy system to solve.

This monograph demonstrates that there *is* a simpler theory of relativity. Step by step, we construct relativistic dynamics in as simple a way as possible and still obtain a theory broad enough to explain all known relativistic phenomena. Our theory is objectively simpler than GR. In GR, the metric for a gravitational field has *ten* degrees of freedom. In our model, it has only *three* degrees of freedom. Our mathematics is also simpler than that of GR. With a background of only multivariable calculus and linear algebra, the reader will be able to follow the derivation of Mercury's precession, the deflection of light, and all of the other tests of GR. Our approach can also be used to derive the Biot-Savart Law and Snell's Law.

1.3 Overview of the Model

Before previewing our model, we should first define the term *relativistic dynamics*. For us, relativistic dynamics is a theory of the motion of objects, influenced by force fields and isotropic media, which is Lorentz invariant, reduces to Newtonian dynamics when velocities are small, and explains all known relativistic effects.

Newtonian dynamics is often a good enough approximation to relativistic dynamics. If the magnitude of the measurement errors are larger than the relativistic corrections, then relativity is not needed. When the measurements are very accurate, however, one needs relativistic dynamics. It is also possible to observe the relativistic correction for motion with extremely high frequency, if these corrections are combined.

We note that some define relativistic dynamics as a theory of the motion of objects whose speed is close to the speed of light c. We reject this definition because, for example, one needs relativistic dynamics to account for the anomalous precession of Mercury, even though Mercury travels along its orbit at less than fifty kilometers per second.

We turn now to an overview of our model. In the earlier chapters, we study the motion of objects, ignoring any internal rotation and consider objects as points. Rotations are handled in the later chapters. Throughout the entire text, however, we treat, primarily, action at a distance.

We use simplicity and choose to describe the motion of objects in an *inertial* reference frame attached to an inertial observer. Our observer describes events in flat spacetime. This is similar to the problem of finding the shortest route on the Earth's surface, where it is easier to do the calculations on a flat map, rather than comparing the length of various routes on the globe itself.

One way to describe the motion of an object is by a function assigning the position of the object at different times. We work, however, with a geometric description – the graph of this function, which is a line, called the *worldline*, in spacetime. A worldline has the general form $x(\sigma) = x^\mu(\sigma)$, where $\mu = 0, 1, 2, 3$. Here, x^0 is the time of the event (or a multiple of the time in order to make x^0 have dimensions of length) in our inertial frame; x^1, x^2, x^3 are the spatial coordinates in this frame, and σ is an arbitrary parameter. Our method of measuring spacetime distances will be shown to be independent of the choice of parameter. This allows us the freedom to choose the parameter of our liking.

The next step is to introduce the *relativity of spacetime*, or the notion of an *object's spacetime*. This means that spacetime is an object-dependent notion. Each object has its own spacetime, which is defined by the forces that affect its motion and at most one parameter intrinsic to the object. A massive, non-charged object's spacetime is affected by gravity but not by electromagnetic forces. Its spacetime is that of the curved spacetime induced by nearby masses. An electron is affected by both gravity and electromagnetism. Its spacetime reflects the combined effect of both forces and depends on its charge-to-mass ratio. Similarly, the spacetimes of photons and charges traveling through a medium are determined by the properties of the medium. A photon's spacetime may depend on its frequency, as we observe when a prism splits white light into a rainbow.

Next, we introduce the *Extended Principle of Inertia*, which states:

Since an inanimate object is unable to change its velocity, it moves via the shortest, or stationary, worldline in its spacetime when not disturbed by other objects.

This principle extends Newton's First Law from free motion in flat spacetime to the motion of an arbitrary object in *its* spacetime. It means that objects move along geodesics, as suggested by Riemann and as in GR. Only now the geodesics are with respect to a geometry (not necessarily defined by a metric) that not only incorporates gravity, but also electromagnetic forces and the influence of media. In GR, an object freely falling in a gravitational field is in free motion. In our dynamics, *every* object is in free motion in *its* spacetime.

To find stationary paths, we use the Principle of Least Action, but with a physically meaningful action. Historically, actions have been defined differently in different areas of physics. Often, there is no physical understanding behind an action's definition. It simply works. Here, on the other hand, we propose a simple, physically meaningful action function that generalizes the Lagrangian but has no connection to the usual "kinetic minus potential energy."

In order to define the "shortest" worldline, we need to define the "length" of a worldline in an object's spacetime. To do this, it is enough to define the distance between two infinitesimally close spacetime points P and Q. This is the analog of the line element of the spacetime metric in GR [59, 78]. We propose the following definition:

Definition 1 The action function $L(x, u)$ is a scalar-valued function of the spacetime position x and a four-vector u, with the meaning that the distance between two spacetime positions $P = x$ and $Q = x + u\epsilon$ in the object's spacetime is $L(x, u)\epsilon$ if ϵ is small.

We think of u as the direction from P to Q. On the worldline of an object, we will substitute its four-velocity for u.

Calculus shows us that to measure the length of a curve, it is enough to know the approximate length of infinitesimal segments of the curve. Sometimes a linear approximation is sufficient; sometimes one needs higher-order approximations. For us, the length is infinitesimal in u alone, and we will see that for the electromagnetic field, linear approximations of u are enough, while quadratic approximations are needed for gravity.

We will show that the simplest such action function $L(x, u)$ that is Lorentz invariant and independent of the parametrization is

$$L(x, u) = \sqrt{u^2 - (l(x) \cdot u)^2} + kA(x) \cdot u, \tag{1.1}$$

where $l(x)$ and $A(x)$ are four-vector-valued functions of the spacetime position x, $a \cdot b = \eta_{\mu\nu} a^\mu b^\nu$ is the Minkowski inner product, and k is a parameter which can be positive or negative. The expression under the square root must be non-negative.

This entails a restriction on the admissible four-velocities in the spacetime under investigation.

We call $A(x)$ the *linear four-potential* because it acts linearly on the four-velocity. We call $l(x)$ the *quadratic four-potential* because it act quadratically on the four-velocity. We will see that the function $kA(x)$ is connected to the electromagnetic field and that $k \sim q/m$, while the function $l(x)$ describes the gravitational field and the effect of an isotropic medium. For gravitational fields, $l(x)$ is a null vector proportional to the direction of propagation of the field. For the motion of charges and light in isotropic media, $l = l(x)$ is timelike and does not depend on the spacetime position x. The exact form of $A(x)$ and $l(x)$ will be obtained using Lorentz covariance and the Newtonian limit.

Stationary worldlines will be obtained by applying the Euler-Lagrange equations (Figs. 1.5 and 1.6) to the above action function. This entails choosing a parameter on the worldline of the motion. We usually use proper time as the parameter, since it is the same for all inertial observers. In some applications, however, we do take advantage of the freedom of choice of the parameter and use a parameter which simplify the computations.

To obtain the general equation of motion for electromagnetism and gravity and motion in isotropic media, we introduce, for any four-vector-valued function f, a first-order derivative

$$F^{\alpha}_{\nu}(f) = \eta^{\alpha\lambda}(f_{\nu,\lambda} - f_{\lambda,\nu}), \tag{1.2}$$

Fig. 1.5 Leonhard Euler (1753 portrait by Jakob Emanuel Handmann)

Fig. 1.6 Joseph-Louis Lagrange

which is the antisymmetric part of the Jacobian matrix of f. If $f = A$, then $F^{\alpha}_{\nu}(A)$ is the usual electromagnetic field strength tensor.

For arbitrary $A(x)$ and $l(x)$, we define

$$b^{\alpha}(A, l) = \sqrt{1-(l\cdot\dot{x})^2}\,kF^{\alpha}_{\nu}(A)\dot{x}^{\nu} - (l\cdot\dot{x})F^{\alpha}_{\nu}(l)\dot{x}^{\nu}, \tag{1.3}$$

where the dot denotes differentiation by the proper time. For $l = 0$, b reduces to the acceleration under the Lorentz force. In general, however, this acceleration must be multiplied by the gravitational time dilation, expressed by the square root term. The second term accounts for the gravitational force, which depends quadratically on the four-velocity.

Applying the Euler-Lagrange equations to the action function (1.1) leads to a universal relativistic dynamics equation

$$\ddot{x} = b + (b\cdot l)l^{\perp} + (\dot{l}\cdot\dot{x})l^{\perp}, \tag{1.4}$$

where $l^{\perp} = l - (l\cdot\dot{x})\dot{x}$. If there is only an electromagnetic field, then only the first term is nonzero, and A is a four-potential of the field. If there is a gravitational field, the first term gives the Newtonian limit ($v \ll c$) for a weak gravitational field. The first two terms, taken together, give the Newtonian limit even in the strong regime. The $l^{\perp}$ terms ensure that an object's four-velocity remains within the admissible region. Just like in the case of an electromagnetic field, we obtain that a gravitational field splits into a near field, falling off like $1/r^2$, and a far field, falling off like $1/r$. The far field is part of the third term. For a pure gravitational field, the dynamics resulting from Eq. (1.4) and conservations following from the action function (1.1) passes all of the tests of GR.

For the motion of light in isotropic media, we use the Lorentz invariance of the action function to obtain the effect of the motion of a medium on the velocity of light in the medium. We also derive Snell's Law for light propagation between two media. Then we derive the equation of motion for charges in an isotropic medium at rest. This equation is essentially (1.4), but with a slight modification. Thus, Eq. (1.4), together with conservation laws, describes the motion of both charged and uncharged, massive objects and massless particles, in any electromagnetic and some gravitational fields in vacuum, as well as in isotropic media.

We point out the advantages of the action function (1.1) over the more standard Lagrangian $L_{standard} = \frac{1}{2}mv^2 - U$, kinetic energy minus potential energy. This Lagrangian depends on the mass m of the object in motion, and yet the motion in a gravitational field is independent of the mass. In other words, the standard Lagrangian contains a parameter which has no bearing on the physics! On the other hand, the gravitational term of our action function (1.1) does not depend on the mass. Moreover, the linear term depends on the charge-to-mass ratio, a parameter on which the acceleration due to an electromagnetic field is known to depend.

In the later chapters, we extend the description of an object from a point to an object with internal rotation. This forces us to complexify Minkowski space. Using

Lorentz covariance, we obtain a relativistic description of spin that predicts the correct transition probabilities between states and leads to a relativistic dynamics equation for particles with spin. In real Minkowski space, as we will show, there are only two types of Lorentz-covariant vectors and one type of Lorentz-invariant scalar. These are needed to describe electromagnetic and gravitational fields. In complex Minkowski space, we discovered that there is an additional complex-valued, Lorentz-invariant scalar associated to any null vector. This allows us to obtain a new relativistic description of any single-source field which propagates with the speed of light. Our description is similar to the wave function in quantum mechanics.

1.4 Outline of the Book

The classical treatment of motion in electromagnetic and gravitational fields is reviewed in Chap. 2. Here, the reader will find explicit solutions for both bounded and unbounded trajectories. We prove Kepler's laws of planetary motion and derive the Euler-Lagrange equations, which play a central role in the book.

Since we rely heavily on the Principle of Relativity and Lorentz covariance, we review these concepts in Chap. 3. We use the symmetry following from the Principle of Relativity to derive the Lorentz transformations. We show that there are a universal speed and a metric which are invariant for all inertial frames. We also derive velocity transformations between inertial frames and demonstrate some of their physical applications, including the Sagnac effect. We discuss energy-momentum of massive and massless particles and use it to derive the relativistic Doppler shift. We also establish here the allowable forms of Lorentz-covariant vector-valued functions.

Although there are no new results in this chapter, one will nevertheless find there:

- a precise and local definition of *inertial frame*,
- a proof that the Galilean transformations and the Lorentz transformations are the only transformations between inertial frames that satisfy the Principle of Relativity,
- a derivation of the Lorentz transformations without assuming that the speed of light is the same in all inertial frames,
- visualizations and geometric explanations of both the Galilean and the Lorentz transformations,
- arguments why the Lorentz, and not the Galilean, transformations are the true transformations between inertial frames,
- a clear explanation of the difference between vectors and covectors, and the need for both,
- the definition of the energy-momentum of massive and massless free particles,
- a derivation of the relativistic Doppler shift based on energy-momentum,
- a characterization of the Lorentz-covariant covectors associated to the field of a single source.

In Chap. 4, we present the foundations of our approach - the relativity of spacetime and the Extended Principle of Inertia. We introduce the action function (1.1) and derive the relativistic dynamics Eq. (1.4). To help the reader familiarize himself with action functions and stationary worldlines, we tackle the ancient problem of finding the shortest route between two points on the Earth's surface. Our approach includes a method of visualizing the geometry by plotting sections of the action function. We distinguish between the momentum of a free particle and the momentum imparted by a field to a particle. We derive the equation of motion with respect to two different parametrizations of the worldline of a moving object.

Chapter 5 is devoted to the electromagnetic field. Using Lorentz covariance and the fact that, in general, electromagnetic fields radiate, we obtain the Liénard-Wiechert four-potential of a single source. We describe all of the properties of the field of a moving source and compute explicitly the near and far fields. Our presentation includes visualizations of the geometry of the fields and a new interpretation of the magnetic field as the angular momentum of the source. We introduce a new definition of the *energy-momentum four-vector of a field*. The connection to Maxwell's equations are explored here as well. We compute relativistic orbits in a static, single-source field.

Gravity is treated in Chap. 6. From Lorentz covariance and the Newtonian limit, we obtain the field of a spherically symmetric source at rest. We present visualizations of the geometry and compare them to those of the electromagnetic field. We note that a *charge* at rest does not impart 3D momentum to test charges, while a *gravitational source* does imparts 3D momentum to test objects, *even when the source is at rest*.

We derive the correct precession of Mercury and show that our model passes all of the other tests of GR, including gravitational waves. The results are extended to the field of a collection of moving, spherically symmetric sources. Circular and elliptical orbits are computed in the strong field regime.

The motion of light and charged particles in isotropic media is handled in Chap. 7. We analyze Fizeau's experiment and derive Snell's Law. In Chap. 8, we complexify Minkowski space and obtain a relativistic description of spin using complex four-vectors, but without using spinors. We derive the transition probabilities between states and a relativistic dynamics equation for particles with spin. Our model agrees with quantum mechanics with regard to Bell's inequality. In Chap. 9, still working in complexified Minkowski space, we find a Lorentz-invariant scalar-valued function which can be used to describe the field of a moving source. This description is Lorentz invariant with respect to a spin-1/2 representation of the Lorentz group and produces a prepotential similar to the wave function of quantum mechanics. This prepotential leads to the Liénard-Wiechert four-potential.

1.5 Alternative Theories

In the literature, there are other alternative approaches to reproducing the relativistic gravitational features of GR. One approach uses modified Newtonian-like potentials. This so-called "pseudo-Newtonian" approach, introduced in [81], is much simpler mathematically than GR, with no need for covariant differentiation and complicated tensorial equations. Numerous authors [1, 16, 58, 69, 72, 75, 88] have proposed various modified Newtonian-like potentials. However, none of these potentials are able to reproduce the tests of GR, even in the weak field regime. Moreover, as stated in [51], most of these modified potentials "are arbitrarily proposed in an ad hoc way" and, more fundamentally, are "not a physical analogue of local gravity and are not based on any robust physical theory and do not satisfy Poisson's equation."

More recently, the above shortcomings were addressed in [52]. Using a metric approach and hypothesizing a generic relativistic gravitational action and a corresponding Lagrangian, the authors derive a velocity-dependent relativistic potential which generalizes the classical Newtonian potential. For a static, spherically symmetric geometry, this potential exactly reproduces relativistic gravitational features, including the tests of GR. Even more recently, one finds a fundamental grounding to these velocity-dependent pseudo-Newtonian potentials in [100]. The authors generalize the pseudo-Newtonian approach to any stationary spacetime. They also include additional forces, such as the electromagnetic force. Dirac [20] combined gravity and electromagnetism into one action function. Mashhoon's [74] gravitoelectromagnetism borrows ideas from electromagnetism to model gravity.

Chapter 2
Classical Dynamics

The two classical fields are the electromagnetic field and the gravitational field. In this chapter, we review the classical dynamics of these inverse-square, central fields and derive the equation of motion for objects or particles moving in these fields. Explicit solutions are obtained for both bounded and unbounded trajectories. We derive Kepler's laws of planetary motion, and, since the Euler-Lagrange equations are central to our model, we provide a rigorous derivation. Our derivation follows spsciteGoldstein and can be found in standard textbooks on classical mechanics.

2.1 Classical Fields

Classical dynamics is based on Newton's Second Law $\mathbf{F} = m\mathbf{a}$, which defines the 3D acceleration vector $\mathbf{a}$ of an object of mass m moving under a force $\mathbf{F}$. This law can be written in the form

$$\frac{d\mathbf{p}}{dt} = \mathbf{F}, \tag{2.1}$$

where $\mathbf{p} = m\mathbf{v}$ is the momentum of the moving object. We consider primarily action at a distance, which can be viewed as motion under the influence of a field.

Let us first consider the electromagnetic field. Let Q be a point charge at rest at the origin. The potential $U^{\mathrm{EM}}(\mathbf{x})$, at the spatial point $\mathbf{x} = (x^1, x^2, x^3)$, of the electric field generated by Q is

$$U^{\mathrm{EM}}(\mathbf{x}) = \frac{1}{4\pi\epsilon_0}\frac{Q}{|\mathbf{x}|}, \tag{2.2}$$

where $|\mathbf{x}| = \sqrt{(x^1)^2 + (x^2)^2 + (x^3)^2}$ is the length of $\mathbf{x}$ and the distance of $\mathbf{x}$ from the origin, and the constant ϵ_0 is the *permittivity of free space*. In the SI system of units, which we use throughout the book,

Y. Friedman and T. Scarr, *A Novel Approach to Relativistic Dynamics*,
Fundamental Theories of Physics 210,
https://doi.org/10.1007/978-3-031-25214-3_2

$$\epsilon_0 \approx 8.85 \times 10^{-12} \frac{\mathrm{C}^2}{\mathrm{N} \cdot \mathrm{m}^2} \tag{2.3}$$

and

$$\frac{1}{4\pi \epsilon_0} \approx 9 \cdot 10^9 \frac{\mathrm{N} \cdot \mathrm{m}^2}{\mathrm{C}^2}. \tag{2.4}$$

The potential energy of a test charge q at $\mathbf{x}$ is $qU^{\mathrm{EM}}(\mathbf{x})$ and reflects the interaction of the external field, generated by Q, and test charge's own field. Note that a charge q satisfying $qQ > 0$ has positive potential energy, since in this case, the force is repulsive. If the source expels the charge q from the field, q will become a free charge with some nonzero velocity. Thus, its total energy will be its rest energy *plus* the kinetic energy of its motion. This kinetic energy came from the field. If $qQ < 0$, then q has negative potential energy, since in this case, one must *add* energy to free it from the field.

The acceleration $\mathbf{a}$ of a test charge q of mass m at rest at the point $\mathbf{x}$ is computed from $\mathbf{a} = -\frac{q}{m}\nabla U^{\mathrm{EM}}$, and we have

$$\mathbf{a} = \frac{q}{m} \frac{1}{4\pi \epsilon_0} \frac{Q\mathbf{x}}{|\mathbf{x}|^3}. \tag{2.5}$$

The associated force $\mathbf{F} = m\mathbf{a}$ is called the *electric force* or the *Coulomb force*.

The second classical field is the gravitational field. Let M be a point mass at rest at the origin. The gravitational potential (per unit mass) U^{G} at a point $\mathbf{x}$ is

$$U^{\mathrm{G}}(\mathbf{x}) = -\frac{GM}{|\mathbf{x}|}, \tag{2.6}$$

where G is *Newton's gravitational constant* –

$$G \approx 6.674 \times 10^{-11} \frac{\mathrm{N} \cdot \mathrm{m}^2}{\mathrm{kg}^2}. \tag{2.7}$$

The potential energy of an object of mass m positioned at $\mathbf{x}$ is mU^{G}. This potential energy is negative because one must *add* energy to escape the field. Tremendous energy must be given to a rocket, for example, to send it into outer space.

The acceleration $\mathbf{a}$ of a test mass m at rest at the point $\mathbf{x}$ is computed from $\mathbf{a} = -\nabla U^{\mathrm{G}}$, and we have

$$\mathbf{a} = -\frac{GM\mathbf{x}}{|\mathbf{x}|^3}. \tag{2.8}$$

The associated force $\mathbf{F} = m\mathbf{a}$ is called the *gravitational force*.

Note that the gravitational *acceleration* does *not* depend on the mass m or any other property of the test mass, while the electric acceleration (2.5) depends on the charge-to-mass ratio q/m. Both fields are inverse-square fields, meaning that they fall off like $1/|\mathbf{x}|^2$. Both fields are also *central*, meaning that the acceleration is along the line connecting the source and the object. On the other hand, comparing the values of the coefficients G and $1/4\pi\epsilon_0$, we see that the electromagnetic force is substantially larger than the force of gravity.

2.2 Motion in the Classical Fields

A force is called *central* if for some parameter σ, the 3D acceleration $d^2\mathbf{r}/d\sigma^2$ of an object in the field is radial, that is, proportional to the 3D position vector $\mathbf{r}$.

Lemma 1 If the acceleration $\ddot{\mathbf{r}}^2 = d^2\mathbf{r}/d\sigma^2$ with respect to a given parameter σ is radial, then the motion is in the plane Π spanned by the initial position $\mathbf{r}$ and the initial velocity $\dot{\mathbf{r}} = d\mathbf{r}/d\sigma$. Orbital angular momentum is conserved. Using polar coordinates r, φ in Π, the orbital angular momentum has constant magnitude

$$l = r^2\dot{\varphi}. \tag{2.9}$$

The square $\dot{\mathbf{r}}^2$ of the magnitude of the velocity is

$$\dot{r}^2 + r^2\dot{\varphi}^2 = \dot{r}^2 + \frac{l^2}{r^2}. \tag{2.10}$$

(The quantity $\dot{\varphi}$ is called the *angular velocity* of the motion.)

Proof Since the acceleration is radial, $\mathbf{r} \times \frac{d^2\mathbf{r}}{d\sigma^2} = 0$. Hence,

$$\frac{d}{d\sigma}\left(\mathbf{r} \times \frac{d\mathbf{r}}{d\sigma}\right) = 0 + \mathbf{r} \times \frac{d^2\mathbf{r}}{d\sigma^2} = 0.$$

This implies that the orbital angular momentum $\mathbf{L} = \mathbf{r} \times m\frac{d\mathbf{r}}{d\sigma}$ is constant. Thus, the plane Π generated by the initial position $\mathbf{r}$ and the initial velocity $d\mathbf{r}/d\sigma$ is perpendicular to the vector $\mathbf{L}$, which remains constant. This implies that the motion remains in the plane Π. Using polar coordinates r, φ in Π, and using a dot to denote differentiation by σ, we have

$$\mathbf{r} = (r\cos\varphi, r\sin\varphi)$$

and

$$\dot{\mathbf{r}} = \dot{r}(\cos\varphi, \sin\varphi) + r\dot{\varphi}(-\sin\varphi, \cos\varphi).$$

Hence, the magnitude of $\dot{\mathbf{r}}^2$ is $\dot{r}^2 + r^2\dot{\varphi}^2$, and the magnitude of the angular momentum $\mathbf{r} \times \dot{\mathbf{r}}$ is $r^2\dot{\varphi}$. This proves the lemma.

We derive now the *classical equation of motion* in an inverse-square, central force field. By assumption, the force $\mathbf{F}(\mathbf{r})$ at the spatial point $\mathbf{r} = (x^1, x^2, x^3)$ has the form

$$\mathbf{F}(\mathbf{r}) = m\frac{d^2\mathbf{r}}{dt^2} = \alpha\frac{\mathbf{r}}{r^3}, \tag{2.11}$$

where $r = \sqrt{(x^1)^2 + (x^2)^2 + (x^3)^2}$ and α is a constant which is positive for repulsive forces and negative for attractive forces. Both the Coulomb force and the gravitational force are of this type (see formulas (2.5) and (2.8)). Moreover, the force is conservative, since $\mathbf{F}(\mathbf{r}) = -\nabla U(\mathbf{r})$, where $U = \frac{\alpha}{r}$. We will parameterize the object's trajectory by ct, where c is the speed of light in vacuum, rather than t, in order to make its velocity unit free.

To derive the equation of motion, we start with conservation of the total energy $E = 1/2mv^2 + U(\mathbf{r})$ of the moving particle. E is the sum of the object's kinetic energy and the potential energy of the field. Thus,

$$\frac{1}{2}mc^2\dot{\mathbf{r}}^2 + \frac{\alpha}{r} = E,$$

where the dot denotes differentiation by ct. This implies that $c^2\dot{\mathbf{r}}^2 = v^2$, where v is the velocity of the particle. Since the acceleration is radial, Lemma 1 holds, and the energy conservation can be written as

$$\frac{1}{2}mc^2\left(\dot{r}^2 + \frac{l^2}{r^2}\right) + \frac{\alpha}{r} = E.$$

Multiply the previous equation by $2/mc^2$ to obtain the *dimensionless energy conservation equation*

$$\dot{r}^2 + \frac{l^2}{r^2} - \frac{r_s}{r} = \mathcal{E}, \tag{2.12}$$

where $r_s = -2\alpha/mc^2$ and $\mathcal{E} = 2E/mc^2$.

We are primarily concerned with bounded trajectories, in which case the force must be attractive. In this case, $\alpha < 0$, and so $r_s > 0$. A particle can escape the attraction of the field only if on the escape trajectory the energy $\mathcal{E} \geq 0$, since it is equal to the kinetic energy at infinity, where the potential energy vanishes. This implies that the initial velocity must satisfy $v^2/c^2 \geq r_s/r$. Thus, if the initial position is in the region $r < r_s$, then the escape velocity of the particle from the field is greater than c, which is impossible. The quantity r_s is called the *Schwarzschild radius*.

Next, introduce

$$f(\varphi) = r_s/r(\varphi), \tag{2.13}$$

which is the *dimensionless potential energy* on the trajectory. Note that $f(\varphi)$ must be positive on the trajectory. Using a prime to denote differentiation by φ, we have

$$f' = \frac{df}{d\varphi} = -\frac{r_s}{r^2}\frac{dr}{d\varphi} = -\frac{r_s}{r^2}\frac{\dot{r}}{\dot{\varphi}} = -\frac{r_s}{l}\dot{r}, \tag{2.14}$$

implying that $\dot{r} = -\frac{l}{r_s} f'$. Eq. (2.12) can thus be rewritten as

$$\frac{l^2}{r_s^2}\left(f'\right)^2 + \frac{l^2}{r_s^2} f^2 - f = \mathcal{E}. \tag{2.15}$$

Define a new dimensionless constant

$$\mu = \frac{r_s^2}{2l^2} \tag{2.16}$$

and rewrite (2.15) as

$$\left(f'\right)^2 = -f^2 + 2\mu f + 2\mu\mathcal{E}. \tag{2.17}$$

Differentiate (2.17) by φ to get

$$2f'f'' + 2ff' - 2\mu f' = 0. \tag{2.18}$$

Divide by $2f'$ to obtain the easily solved, second-order differential equation

$$f'' + f = \mu. \tag{2.19}$$

The general solution is

$$f(\varphi) = \mu(1 + e\cos(\varphi - \varphi_0)), \tag{2.20}$$

where e is a constant and φ_0 is the polar angle of the *periapsis* – the point of the orbit nearest the source. From (2.13), the orbit is given by

$$r(\varphi) = \frac{r_s/\mu}{1 + e\cos(\varphi - \varphi_0)}. \tag{2.21}$$

Therefore, trajectories in an inverse-square, central force field have the form (2.21), with the condition that $r(\varphi) > 0$.

We consider now the two cases $0 < e < 1$ and $e > 1$.

Case 1 $0 < e < 1$

In this case, it follows from Eq. (2.20) that $f(\varphi)$ is positive for all values of φ. This implies that the trajectory exists for all values of φ. The function is periodic with period 2π and has both a maximum and a minimum.

The maximal value

$$f_p = \mu(1 + e) \tag{2.22}$$

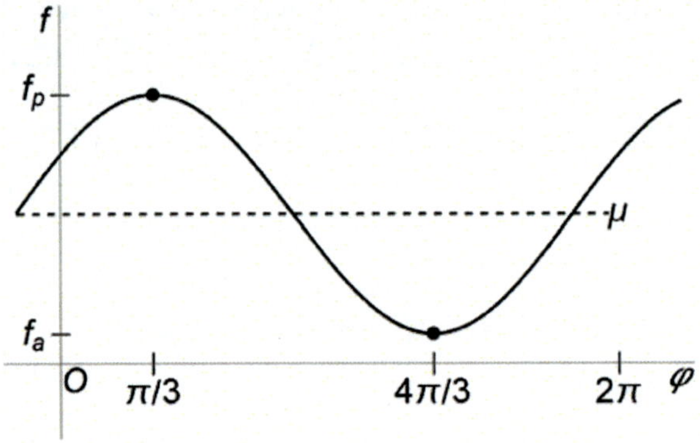

Fig. 2.1 The function $f(\varphi)$ is the negative of the potential energy and has the form $f(\varphi) = \mu(1 + e\cos(\varphi - \varphi_0))$. In the figure, $\mu = 1.5 \times 10^{-5}$, $e = 0.8$ and $\varphi_0 = \pi/3$. The values of $f(\varphi)$ vary between f_a and f_p, and $f(\varphi) > 0$ for all values of φ. The corresponding orbit is bounded

of $f(\varphi)$ obtains at $\varphi = \varphi_0$ and corresponds to the minimal value r_p of $r(\varphi)$. This occurs at the periapsis. The minimal value

$$f_a = \mu(1 - e) \tag{2.23}$$

of $f(\varphi)$ obtains at $\varphi = \varphi_0 \pm \pi$ and corresponds to the maximal value r_a of $r(\varphi)$. This occurs at the *apoapsis* - the point of the orbit furthest from the source. The polar angle $\varphi_0 \pm \pi$ is the *angle of apoapsis*. Clearly, $\dot{r}(r_p) = \dot{r}(r_a) = 0$. See Fig. 2.1.

The solution (2.21) for a particular orbit is completely determined by the measurable quantities r_s and φ_0, along with the initial conditions $f(\varphi_0) = f_p$ and $f(\varphi_0 \pm \pi) = f_a$. To compute μ and e, use (2.22) and (2.23):

$$\mu = \frac{f_p + f_a}{2} \quad , \quad e = \frac{f_p - f_a}{2\mu}.$$

We show now that the trajectory (2.21) is an elliptical orbit, with eccentricity e and semi-latus rectum r_s/μ. Since the minima of $r(\varphi)$, corresponding to the periapsis, occur when $\varphi = \varphi_0 + 2\pi n$, $n = 0, 1, 2, \cdots$, the position of the periapsis will not change with the revolution of the object. This means that this elliptical orbit *does not precess*.

We show, in fact, that Eq. (2.21) defines an ellipse with one focus F_1 at the origin and a second focus F_2 at $(-2c, 0)$, for some positive number c which depends on the eccentricity e of the ellipse (see Fig. 2.2). A point A is on this ellipse if and only if its total distance to the foci $|F_1A| + |AF_2|$ is $2a$, for some $a > 0$. Let A be an arbitrary point, with polar coordinates r, φ. By the Law of Cosines, $AF_2 = \sqrt{r^2 + 4c^2 + 4rc\cos\varphi}$. The equation of the ellipse is

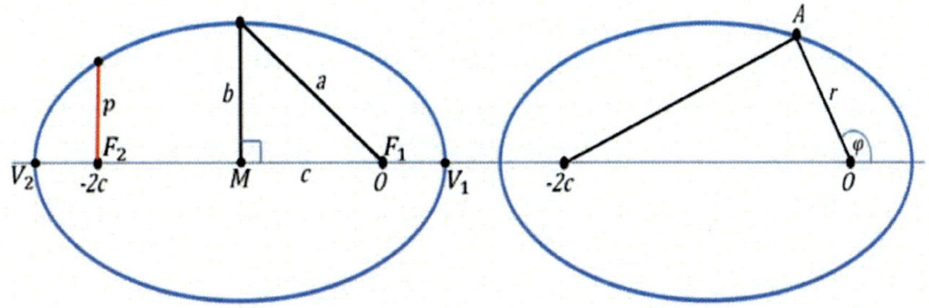

Fig. 2.2 (Left) An ellipse and its parameters. The focus F_1 is at the origin, and the focus F_2 is at $(-2c, 0)$. The line segment connecting the two vertices V_1 and V_2 is called the major axis and has length $2a$. The vertical distance from the center M of the ellipse to the ellipse is the semi-minor axis and has length b, where $c^2 = a^2 - b^2$. The eccentricity of the ellipse is $e = c/a < 1$. The semi-latus rectum $p = b^2/a$ (shown in red) is the vertical distance from either focus of the ellipse to the ellipse. (Right) The sum of the distances from any point on the ellipse to the two foci is $2a$

$$r + \sqrt{r^2 + 4c^2 + 4rc\cos\varphi} = 2a,$$

or

$$\sqrt{r^2 + 4c^2 + 4rc\cos\varphi} = 2a - r.$$

Squaring both sides and canceling, we get

$$r = \frac{a^2 - c^2}{a + c\cos\varphi},$$

which is (2.21) with $r_s/\mu = (a^2 - c^2)/a$ and $e = c/a$.

Kepler's laws of planetary motion characterize bound orbits in a single-source gravitational field, when the source is at rest spscitearnold. We derive these laws now.

Kepler's three laws are:

1. The orbit of the object is an ellipse with the source at one of the two foci.
2. A line segment joining the object and the source sweeps out equal areas during equal intervals of time.
3. The square of the orbital period of the object is directly proportional to the cube of the semi-major axis of its orbit.

First, in the notation of Fig. 2.2, the Pythagorean Theorem implies that $b^2 = a^2 - c^2$, which implies that

$$b^2 = \frac{ar_s}{\mu}. \tag{2.24}$$

We will need this formula shortly.

We have, in fact, established (1) above. For (2), the area $A(\Delta t)$ swept out in a time interval Δt is $\Delta A = (1/2)r^2\Delta\varphi$. Thus, by conservation of angular momentum,

$$\frac{dA}{cdt} = \frac{1}{2}r^2\dot{\varphi} = \frac{1}{2}l$$

is constant. Assume without loss of generality that $A(0) = 0$. Integrating, we have

$$A(t) = \frac{1}{2}clt. \tag{2.25}$$

This proves (2). Note that if $t = T$, the period of the orbit, then formula (2.25) says that the area of the ellipse is $(1/2)clT$. Thus,

$$\pi ab = \frac{clT}{2},$$

and

$$T^2 = \frac{4\pi^2a^2b^2}{c^2l^2}.$$

Now in a gravitational field, the α of Eq. (2.11) is $\alpha = -GMm$ (see formula (2.6) and the ensuing discussion). Hence, in this case, $r_s = -2\alpha/mc^2 = 2GM/c^2$. Thus, using (2.16) and (2.24), we have

$$\frac{b^2}{l^2} = \frac{ar_s}{\mu}\cdot\frac{2\mu}{r_s^2} = \frac{ac^2}{GM}.$$

Substituting this in the above expression for T^2 yields

$$T^2 = \frac{4\pi^2}{GM}a^3. \tag{2.26}$$

This proves (3).

Case 2 $e > 1$

In this case, $f(\varphi) > 0$ only for $\cos(\varphi - \varphi_0) > -1/e$, or $\varphi_0 - \arccos(-1/e) < \varphi < \varphi_0 + \arccos(-1/e)$. This implies that the orbit is unbounded. It has a maximal value given by (2.22) which obtains at the periapsis. There is no apoapsis, as the minimal value of f is negative. See Fig. 2.3.

We show, in fact, that Eq. (2.21) defines a hyperbola with one focus at the origin and a second focus at $(2c, 0)$, for some positive number c which depends on the eccentricity e of the ellipse (see Fig. 2.4).

A point A is on this hyperbola if and only if the difference of the distances to the foci $|F_2A| - |AF_1|$ is $2a$, for some $0 < a < c$. Let A be an arbitrary point, with

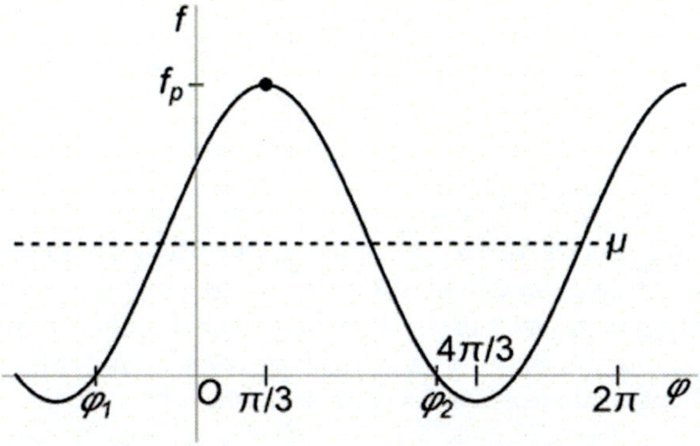

Fig. 2.3 The function $f(\varphi)$ is the negative of the potential energy and has the form $f(\varphi) = \mu(1 + e\cos(\varphi - \varphi_0))$. In the figure, $\mu = 1.5 \times 10^{-5}$, $e = 1.2$ and $\varphi_0 = \pi/3$. $f(\varphi) > 0$ only for $\varphi_1 < \varphi < \varphi_2$, where $\varphi_1 = \varphi_0 - \arccos(-1/e)$ and $\varphi_2 = \varphi_0 + \arccos(-1/e)$. The corresponding orbit is unbounded

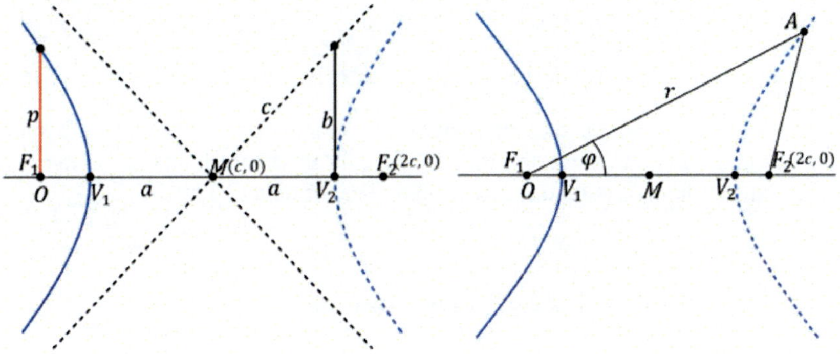

Fig. 2.4 (Left) A hyperbola and its parameters. The focus F_1 is at the origin, and the focus F_2 is at $(2c, 0)$. The line segment connecting the two vertices V_1 and V_2 is called the major axis and has length $2a$. The straight dashed lines are the asymptotes. They intersect at M and have slope $\pm b/a$. The vertical distance from a vertex of the hyperbola to the hyperbola is the semi-minor axis and has length b, where $c^2 = a^2 - b^2$. The eccentricity of the hyperbola is $e = c/a > 1$. The semi-latus rectum $p = b^2/a$ (shown in red) is the vertical distance from either focus of the hyperbola to the hyperbola. (Right) The absolute value of the difference of the distances from any point on the hyperbola to the two foci is $2a$

polar coordinates r, φ. By the Law of Cosines, $AF_2 = \sqrt{r^2 + 4c^2 - 4rc\cos\varphi}$. The equation of the hyperbola is

$$\sqrt{r^2 + 4c^2 - 4rc\cos\varphi} - r = 2a,$$

or

$$\sqrt{r^2 + 4c^2 - 4rc\cos\varphi} = 2a + r.$$

Squaring both sides and canceling, we get

$$r = \frac{c^2 - a^2}{a + c\cos\varphi},$$

which is (2.21) with $r_s/\mu = (c^2 - a^2)/a$ and $e = c/a > 1$.

The trajectory is a hyperbola which is completely determined by the measurable quantities r_s, φ_0, and the unit-free velocity $v(\varphi_0) = v_p$, along with the initial conditions $f(\varphi_0) = f_p$ and $f'(\varphi_0) = 0$. At the periapsis, we have $\dot{r} = 0$, and so $v_p = r_p\dot{\varphi}_p$, where $r_p = r_s/f_p$ and $\dot{\varphi}_p$ is the value of $d\varphi/cdt$ at the periapsis. Thus, using (2.9), the angular momentum, which is constant, has the value $l = r_p v_p$. Hence, from (2.16),

$$\mu = \frac{1}{2}\left(\frac{r_s}{r_p v_p}\right)^2. \tag{2.27}$$

Now use the previous equation and (2.22) to compute

$$e = \frac{2r_p v_p^2}{r_s} - 1. \tag{2.28}$$

The above formula, together with $e > 1$, implies that $v_p^2 > r_s/r_p$. This means that the object can escape the field if and only if its kinetic energy at the periapsis is greater than its potential energy there. See formula (2.12).

2.3 The Euler–Lagrange Equations

In our geometric approach to relativistic dynamics, objects move along stationary worldlines, or *geodesics*. These are worldlines with the shortest distance between any two given points on the line. We therefore need a method to measure distances and identify stationary worldlines, in flat spacetime as well as in curved spacetimes. For this, we use the Calculus of Variations to derive the well-known *Euler-Lagrange equations* (see, for example, spsciteGoldstein). These are a set of second-order differential equations, and for each initial spacetime position and initial velocity, they have a unique solution. Thus, the Euler-Lagrange equations turn the global problem of finding the shortest path between two distant points into a local problem. The Euler-Lagrange equations are valid in any number of dimensions. We will derive them here in four dimensions, since we work in four-dimensional spacetimes throughout the book.

Let $L(x, u)$ be a function of eight variables x^μ, u^μ, for $\mu = 0, 1, 2, 3$, where x is a spacetime position and u is a four-vector (for the precise definition of *four-vector*, see sect. 3.1). Our notation uses upper indices for x and u, while their derivatives will be given lower indices. The rationale for this will be explained in the next chapter.

For an arbitrary worldline $x = x^\mu(\sigma), a \le \sigma \le b$, define the *length* or *action* $S[x(\sigma)]$ of the worldline to be

$$S[x(\sigma)] = \int_a^b L\left(x(\sigma), \frac{dx(\sigma)}{d\sigma}\right) d\sigma\,. \tag{2.29}$$

The worldline $x(\sigma)$ is called *stationary* if

$$\frac{d}{d\epsilon} S[x(\sigma) + \epsilon h(\sigma)]\Big|_{\epsilon=0} = 0 \tag{2.30}$$

for any smooth $h : [a, b] \to h(\sigma)$ satisfying $h(a) = h(b) = 0$.

For each μ, define the *unit-free energy-momentum covector associated with* $L(x, u)$ as

$$p_\mu(\sigma) = \frac{\partial L(x, u)}{\partial u^\mu}\Big|_{x=x(\sigma),\, u=dx(\sigma)/d\sigma}. \tag{2.31}$$

Theorem *The worldline $x(\sigma)$ is stationary if and only if for every μ, we have*

$$\frac{d}{d\sigma} p_\mu(\sigma) = \frac{\partial L(x, u)}{\partial x^\mu}\Big|_{x=x(\sigma),\, u=dx(\sigma)/d\sigma}. \tag{2.32}$$

Equations (2.32) are known as the Euler-Lagrange equations. *The notation in (2.31) and (2.32) means that one first differentiates L by either x^μ or u^μ and then substitutes $x(\sigma)$ for x and $dx(\sigma)/d\sigma$ for u. The Euler-Lagrange equations are the relativistic form of Newton's Second Law (2.1). We call $\partial L(x, u)/\partial x^\mu$ the* four-force.

(Corollary) The Law of Conservation If the function $L(x, u)$ does not depend on x^μ, then the μ component p_μ of the energy-momentum covector, defined by (2.31), is conserved along stationary worldlines:

$$\frac{\partial L}{\partial x^\mu} = 0 \;\Rightarrow\; p_\mu \text{ is constant.} \tag{2.33}$$

Proof of the Theorem First, expand $L\left(x(\sigma) + \epsilon h(\sigma), \frac{d}{d\sigma}(x(\sigma) + \epsilon h(\sigma))\right)$ in a Taylor Series about $(x, u) = (x(\sigma), dx(\sigma)/d\sigma)$:

$$L\left(x(\sigma) + \epsilon h(\sigma), \frac{d}{d\sigma}(x(\sigma) + \epsilon h(\sigma))\right)$$

$$= L\left(x(\sigma), \frac{dx(\sigma)}{d\sigma}\right) + \epsilon h^\mu(\sigma)\frac{\partial L(x, u)}{\partial x^\mu}\Big|_{x=x(\sigma),\, u=dx(\sigma)/d\sigma} + \epsilon\frac{dh^\mu(\sigma)}{d\sigma} p_\mu.$$

We have discarded terms of order ϵ^2 and higher because they will vanish when we differentiate by ϵ at $\epsilon = 0$.

Thus,

$$\int_a^b L\left(x(\sigma) + \epsilon h(\sigma), \frac{d}{d\sigma}(x(\sigma) + \epsilon h(\sigma))\right) d\sigma - \int_a^b L\left(x(\sigma), \frac{dx(\sigma)}{d\sigma}\right) d\sigma$$

$$= \epsilon \int_a^b h^\mu(\sigma) \frac{\partial L(x,u)}{\partial x^\mu}\bigg|_{x=x(\sigma),\, u=dx(\sigma)/d\sigma} d\sigma + \epsilon \int_a^b \frac{dh^\mu(\sigma)}{d\sigma} p_\mu d\sigma. \tag{2.34}$$

Let

$$J = \int_a^b \frac{dh^\mu(\sigma)}{d\sigma} p_\mu d\sigma. \tag{2.35}$$

We will compute J using integration by parts and the fact that h vanishes at the end points a and b. Let

$$f^\mu(\sigma) = p_\mu \quad , \quad dg^\mu = \frac{dh^\mu(\sigma)}{d\sigma} d\sigma.$$

Then

$$df^\mu = \frac{dp_\mu}{d\sigma} \quad , \quad g^\mu(\sigma) = h^\mu(\sigma),$$

so

$$J = -\int_a^b h^\mu(\sigma) \frac{dp_\mu}{d\sigma} d\sigma. \tag{2.36}$$

Dividing (2.34) by ϵ, letting ϵ go to 0, and using (2.36), we get

$$\frac{d}{d\epsilon} S[x(\sigma) + \epsilon h(\sigma)]\bigg|_{\epsilon=0} = \frac{d}{d\epsilon}\left[\int_a^b L\left(x(\sigma) + \epsilon h(\sigma), \frac{d}{d\sigma}(x(\sigma) + \epsilon h(\sigma))\right) d\sigma\right]\bigg|_{\epsilon=0}$$

$$= \int_a^b h^\mu(\sigma)\left(\frac{\partial L(x,u)}{\partial x^\mu}\bigg|_{x=x(\sigma),\, u=dx(\sigma)/d\sigma} - \frac{dp_\mu}{d\sigma}\right) d\sigma.$$

Now, if the worldline $x(\sigma)$ is stationary, then, since $h(\sigma)$ is arbitrary, we have

$$\frac{\partial L(x,u)}{\partial x^\mu}\bigg|_{x=x(\sigma),\, u=dx(\sigma)/d\sigma} = \frac{dp_\mu}{d\sigma} \tag{2.37}$$

for every μ. This completes the proof of the theorem.

Chapter 3
The Lorentz Transformations and Minkowski Space

Based on this century of experience, it is generally supposed that a Final theory will rest on principles of symmetry. - Steven Weinberg

In 1632, in his *Dialogue Concerning the Two World Systems* [49], Galileo Galilei laid down one of the most important symmetries in physics—the *Principle of Relativity*. This is the principle that physical laws must be the same in all inertial frames. This is actually two statements in one:

1. The same experiment observed from different inertial frames will lead to the same physical laws;
2. The same experiment performed in different inertial frames will have the same results.

By *inertial frame*, we mean a frame of reference in which an accelerometer and a gyroscope at rest both measure zero. See the next section for a precise definition.

Interpretation (1) says that physical laws are the same for all observers at rest in an inertial frame. From one and the same experiment, all inertial observers must arrive at the same physical law. To verify that this is the case, we must make sure that our *passive* spacetime transformations satisfy the Principle of Relativity. These transformations translate the spacetime coordinates of an event in one inertial frame to the corresponding coordinates *of the same event* in another frame. Thus, interpretation (1) says that physical laws are invariant under passive transformations. Interpretation (2) says that if two identical experiments are performed in two different inertial frames, the results will be the same. Thus, physical laws must also be invariant under active transformations. Now mathematically, active and passive transformations are indistinguishable, and the interpretation, active or passive, is a matter of choice. Therefore, without loss of generality, we always work with passive transformations.

Y. Friedman and T. Scarr, *A Novel Approach to Relativistic Dynamics*,
Fundamental Theories of Physics 210,
https://doi.org/10.1007/978-3-031-25214-3_3

The transformations used in Newtonian dynamics are the Galilean transformations. Under these transformations, the acceleration of an object is the same in all inertial frames, and the laws of this dynamics are statements about accelerations. Newtonian dynamics is a very successful theory and is used even today in most technologies.

In the early twentieth century, however, Einstein [27–29] recognized that the group of Galilean transformations are only an approximation to the group of Poincaré transformations [77]—the true spacetime transformations between inertial frames. Under the Poincaré transformations, Maxwell's equations are invariant, and the speed of light is the same in all inertial frames. These transformations also explain the dragging of light in moving water and other non-classical phenomena. The linear subgroup of the Poincaré group is called the *Lorentz group*, and a theory is traditionally called *Lorentz covariant* if it satisfies the Principle of Relativity with respect to the Poincaré transformations.

In this chapter, we use the symmetry following from the Principle of Relativity to derive the Lorentz transformations. We show that there are a universal speed and a metric which are invariant for all inertial frames. We also derive velocity transformations and demonstrate some of their physical applications.

Our derivation of the Lorentz transformations does *not* assume a priori that the speed of light is the same in all inertial frames. Since its inception, there have been derivations of Special Relativity which did not make this assumption. The first was by Ignatowsky [94] in 1910. Many others followed; see [6] for a full list of references. Nevertheless, the approach here, based on symmetry, is new and first appeared in [33].

With the Lorentz transformations in hand, we explore some of their consequences, including Einstein velocity addition and the Sagnac effect. We then define Minkowski space (flat spacetime) and develop the machinery necessary to guarantee that our model is Lorentz covariant.

We include a section on relativistic energy-momentum. The basic definitions which appear here can be found in standard relativity textbooks, such as [87]. We apply our model to elastic and completely inelastic collisions as well as to the Doppler effect. Throughout the text, our treatment of energy-momentum plays a central role.

3.1 Inertial Frames

A *frame of reference* K is a four-dimensional coordinate system attached to an observer who is at rest in this system. We refer to the observer as an *inertial observer*. The frame is used to label *events*, that is, single moments in space and time. An event might be a flash of light in a laboratory or the absorption of a photon by an electron in intergalactic space. We take for granted that to each event there corresponds a unique 4-tuple $x = (x^0, x^1, x^2, x^3)$, where x^0 is the time of the event (or a multiple of the time in order to make x^0 have dimensions of length) in K, and $\mathbf{x} = (x^1, x^2, x^3)$ is the spatial position of the event in K (see Fig. 3.1).

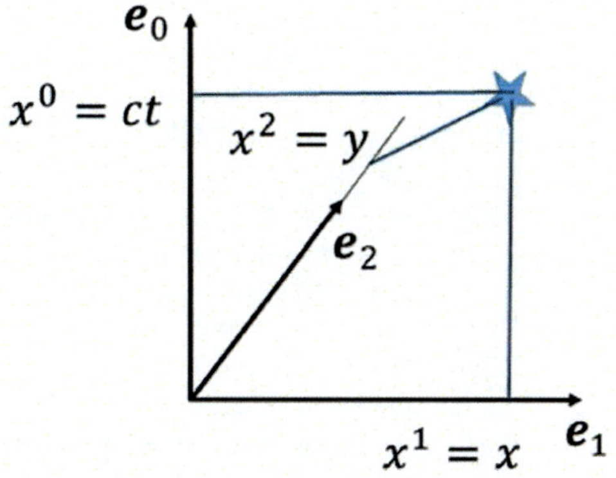

Fig. 3.1 The coordinates of an event in a frame of reference. Three of the four orthonormal axes are shown: the time axis $\mathbf{e}_0$ and two spatial axes $\mathbf{e}_1$ and $\mathbf{e}_2$. The coordinates of the event are found by dropping perpendiculars to each axis. The time is multiplied by the velocity of light c in order to make all coordinates have dimensions of length

We equip our inertial observer with two devices, an accelerometer and a gyroscope. The accelerometer measures the linear acceleration of a rest point of the frame, and the gyroscope measures its rotation. The observer checks these devices. If both the accelerometer and the gyroscope measure zero acceleration, then our observer may assume that no forces are acting on him and that he is an inertial observer. See Fig. 3.2 for a schematic description of an accelerometer. Modern gyroscopes use fiber optics and rely on the Sagnac effect. See Sect. 3.3.2 for details.

Note that, in practice, inertial frames *do not exist* because any nearby masses introduce gravitational forces within the system. Nevertheless, in certain cases, the deviation from an *ideal inertial frame* is small. By the *accuracy* of the frame, we mean the maximum acceleration measured at a rest point of the frame. It is a quantitative measure of how close an inertial frame is to an ideal inertial frame. If the experimental

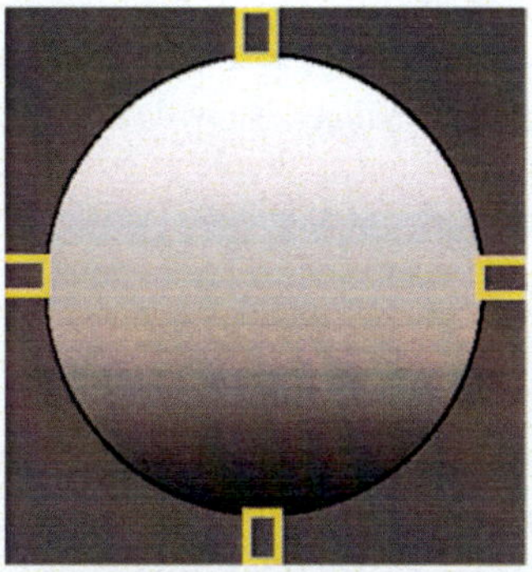

Fig. 3.2 An accelerometer may be thought of as a cubical box containing a small ball at its center. The ball is connected by a sensitive spring in each of three orthogonal directions to an inside wall of the box. If the ball is subject to any force, it will cause the appropriate spring to contract or stretch. Hooke's Law is then used to compute the force on the ball from the spring displacements in each direction. From this force and the known mass of the ball, one calculates the acceleration

accuracy one requires in his experiments inside his frame doesn't reach the accuracy of the frame, then one may consider his frame to be inertial. If more accuracy is required, one could decrease the dimensions of the laboratory. From this point on, therefore, we assume that the inertial frames we use are ideal inertial frames.

Examples of *approximately* inertial frames include a space probe drifting through empty space far away from any massive objects, a satellite orbiting the Earth with the propulsion turned off, a cannonball after being shot from a cannon, an object in a drop tower, and Einstein's elevator falling towards the Earth. In such a frame, Newton's First Law holds approximately. This is because all rest points of the frame have approximately the same acceleration with respect to an outside observer. Therefore, their acceleration relative to each other is approximately zero. We assume that any tidal forces, that is, forces due to the differences in the strength of the field throughout the frame, are negligible. Indeed, inside one of the above frames, one doesn't feel any acceleration. One experiences *near* weightlessness. Aircraft in free fall are used to train astronauts for the experience of weightlessness.

Having defined inertial frames, the next step is to derive the spacetime transformations between two given inertial frames, K and K'. This will enable us to translate the motion of an object as observed in K to its motion as observed in K'. One way to describe the motion of an object is by a function $\mathbf{x}(t) = (x^1(t), x^2(t), x^3(t))$ denoting the spatial position of the object at time t. However, considering only three-dimensional trajectories of motion conceals important information. For example, the same straight line in $\mathbb{R}^3$ can represent both an object moving with constant velocity and an object which is accelerating. It is better to work in four-dimensional *spacetime* and to represent an event using all four of its spacetime coordinates (x^0, x^1, x^2, x^3), where x^0 is the time of the event (or a multiple of the time in order to make x^0 have dimensions of length). In other words, we describe motion using the four-dimensional *graph* of $\mathbf{x}(t)$. In this geometric approach, the motion of an object is described by its *worldline* $x(\sigma) = x^\mu(\sigma)$, where $\mu = 0, 1, 2, 3$. Here, σ is an arbitrary parameter. In what follows, we will show that the dynamics resulting from our model is independent of the choice of parameter. We are then free to choose the parameter of our liking. For the sake of simplicity, we will choose the parametrization that makes the resulting equation of motion as simple as possible.

Our task now is derive the *spacetime transformations* between two arbitrary inertial frames K and K' (see Fig. 3.3). We will show in Sect. 3.2 that there are only two families of spacetime transformations that satisfy the Principle of Relativity. These are the Galilean transformations, used in Newtonian mechanics, and the Poincaré transformations of Special Relativity.

Before we derive spacetime transformations between inertial frames, it is important to distinguish between spacetime points and four-vectors. A *spacetime point* is a 4-tuple $x^\mu = (x^0, x^1, x^2, x^3)$ representing the spacetime coordinates of an event. A *translation* is a function T which acts on spacetime points and has the form

$$T(x^\mu) = x^\mu + b^\mu, \tag{3.1}$$

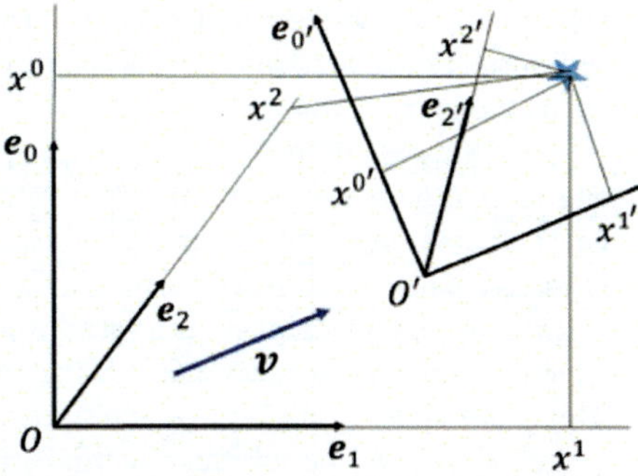

Fig. 3.3 An event as observed in two inertial frames. The spacetime transformation between these frames maps the coordinates of an event in one frame to the coordinates of the same event in the other frame

where b^μ are constants, not all 0. For any spacetime point x^μ and any translation T, we clearly have $T(x^\mu) \neq x^\mu$. That is, spacetime points are not invariant under translations.

On the other hand, the *separation* $x - y$ of two spacetime points x and y *is* invariant under translations, since

$$T(x) - T(y) = (x + b) - (y + b) = x - y. \tag{3.2}$$

Thus, separation of spacetime points is an example of a *four-vector*, a quantity which is invariant under translations. Another example of a four-vector is the four-velocity, defined as follows. Let $x^\mu(\sigma)$ be the worldline of a moving object. The object's *four-velocity with respect to* σ, denoted $u^\mu(\sigma)$, is defined by

$$u^\mu(\sigma) = \frac{dx^\mu(\sigma)}{d\sigma} = \lim_{\varepsilon \to 0} \frac{x^\mu(\sigma + \varepsilon) - x^\mu(\sigma)}{\varepsilon}. \tag{3.3}$$

Since the four-velocity is a limit of four-vectors, it is also a four-vector. The *four-acceleration with respect to* σ, denoted $a^\mu(\sigma)$ and defined by $a^\mu(\sigma) = \frac{d}{d\sigma} u^\mu(\sigma)$, is also a four-vector.

3.2 Spacetime Transformations that Satisfy the Principle of Relativity

In this section, we derive all transformations between inertial frames that satisfy the Principle of Relativity. Until the early twentieth century, the *Galilean transformations*, which are used in Newtonian mechanics, were thought to be the true transformations between inertial frames. However, the experiments of Fizeau [32] and of Michelson and Morley [76] showed that the Galilean transformations are only approximations to the true transformations. Only when the velocity between the frames is small compared to the speed of light is it practical to rely on the Galilean transformations. The true transformations are the *Poincaré transformations*, which we derive here.

The following derivation of the Lorentz transformations is rigorous and includes an explanation of how to bring two inertial frames into *standard configuration* (see below). More importantly, our use of a symmetry of the Principle of Relativity allows us to derive the transformations *without assuming the constancy of the speed of light*. Instead, we obtain that there is a speed which is the same in all inertial frames. From experimental evidence, we conclude that this speed is c, the speed of light in vacuum. This proof appears in [33]. An improved version can be found in [45].

Let K and K' be arbitrary inertial frames. We assume that each frame has a right-handed orthonormal spatial basis, that both frames use the same devices to measure spatial distances, and that the clocks in each frame run at the same rate. Let $L : K' \to K$ be the spacetime transformation mapping the coordinates in K' of an event to the coordinates in K of the same event. We first note that the Principle of Relativity alone implies that L is an *affine* transformation, meaning that L maps lines to lines. To see this, recall that Newton's First Law states that in an inertial frame, an object moves with constant velocity unless acted upon by a force. Such an object is said to be in *uniform* or *free motion*. The worldline of an object in uniform motion is a straight line $x(t) = x(0) + (t, \mathbf{v}t)$, for some constant velocity $\mathbf{v}$. Conversely, if the worldline of a motion is a straight line, then the motion is uniform. Consider now an object moving uniformly in K. By the Principle of Relativity, this object's motion is also uniform in K'. Hence, the worldline in K' of this object is also a straight line. Thus, L maps lines to lines and is an affine transformation. A well-known theorem from linear algebra states that if an affine transformation sends 0 to 0, then the transformation is linear.

We now derive all possible spacetime transformations $L : K' \to K$ which satisfy the Principle of Relativity. We reduce the derivation of the transformations to the case when K and K' are in *standard configuration*. Below we show how to handle the general case. The frames K and K' are said to be in *standard configuration* (see Fig. 3.4) if

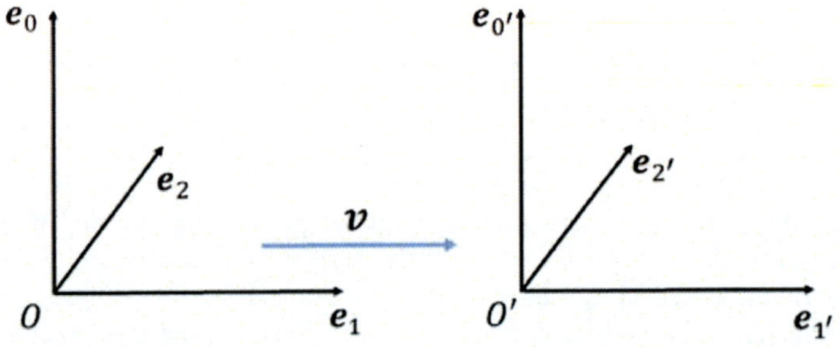

Fig. 3.4 The inertial frames K (on the left) and K' (on the right) are in standard configuration. The velocity of K' in K is $(v, 0, 0)$, while the velocity of K in K' is $(-v, 0, 0)$

1. the origins O and O' coincide at time $t = t' = 0$;
2. the velocity of K' in K is a constant $\mathbf{v} = (v, 0, 0)$, $v \neq 0$, in the positive x direction;
3. the x and x' axes, the y and y' axes, and the z and z' axes, respectively, are parallel.

If K and K' are not in standard configuration, one proceeds as follows. Perform an event at the origin O of K at time $t = 0$ in K. Let $(t'_0, \mathbf{x}'_0)$ be the coordinates of this event in K'. Define an affine translation T by $t'' = t' - t'_0$, $\mathbf{x}'' = \mathbf{x}' - \mathbf{x}'_0$. This is equivalent to moving the origin O' of K' to O and resetting the K' clock at the origin to $t' = 0$. The frames K and (the translated) K' now satisfy property 1 of standard configuration. The worldline of O' in K is now $(t, \mathbf{v}t)$ for some velocity $\mathbf{v}$. Next, perform a rotation R_2 of K' to make $\mathbf{v}$ parallel to the positive x' axis of K', and a rotation R_1 of K to make $\mathbf{v}$ parallel to the positive x axis of K. The resulting frames satisfy property 2 of standard configuration. At this point, there are two possibilities. The first possibility is that the frames now have the same orientation, meaning that a rotation R_3 of the primed frame about the common x, x' axis is sufficient to achieve property 3. In this case, let $R_4 = R_3 R_2$, which is also a rotation. The second possibility is that the frames are oppositely oriented, meaning that the rotation R_3 about the common x, x' axis which makes the y and y' axes parallel makes the z and z' axes antiparallel. In this case, first apply the rotation R_3. Next, apply *space reversal* N to the primed frame. This linear operator takes $t' \to t'$, $x' \to -x'$, $y' \to -y'$, $z' \to -z'$. Now the z and z' axes are parallel, and a final rotation R_5 of the primed frame about the common z, z' will achieve property 3. In this case, let $R_4 = R_5 N R_3 R_2$.

The above procedure factors the transformation $\Lambda : K' \to K$ as

$$\Lambda = R_1^{-1} \Lambda_s R_4 T = \Lambda_l T, \tag{3.4}$$

where Λ_s is the transformation between frames in standard configuration and $\Lambda_l = R_1^{-1} \Lambda_s R_4$ is called the *linear part* of Λ. The transformation Λ is affine, and not linear, due to the translation T. We point out that the general transformation (3.4) and the linear part Λ_l agree on four-vectors x, since by (3.2),

$$\Lambda x = \Lambda_l T x = \Lambda_l x. \tag{3.5}$$

However, for spacetime points x, we have $Tx \neq x$, and hence, in general, $\Lambda x \neq \Lambda_l x$.

We assume for the time being that K and K' are in standard configuration. Notice the lack of symmetry in this configuration (see Fig. 3.4). The velocity of K' in K is $(v, 0, 0)$, while the velocity of K in K' is $(-v, 0, 0)$. To achieve additional symmetry, we perform a reflection sending the x' axis of K' to $-x'$ and denote the new frame by $\tilde{K}'$. This reflection can be achieved by applying space reversal N followed by a rotation. The frames are now in *symmetric configuration* (see Fig. 3.5). Note that the velocity of K in $\tilde{K}'$ is now $(v, 0, 0)$.

We now derive all possible spacetime transformations $\tilde{\Lambda}_s : \tilde{K}' \to K$, where $\tilde{\Lambda}_s$ satisfies the Principle of Relativity, and K and $\tilde{K}'$ are in symmetric configuration. The

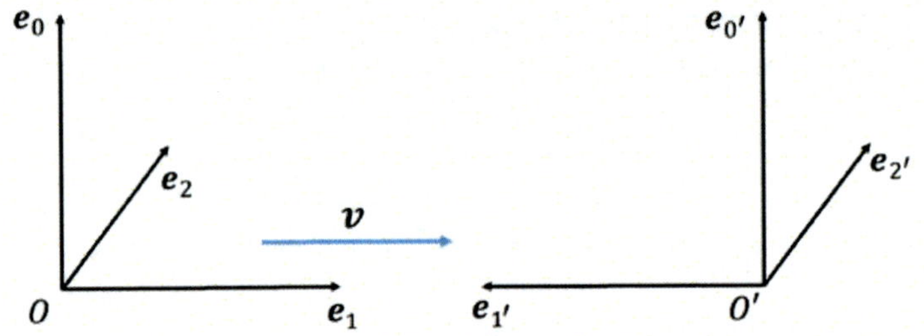

Fig. 3.5 The frames K and $\tilde{K}'$ are in symmetric configuration. Each frame has velocity $(v, 0, 0)$ in the other frame

transformation $\tilde{\Lambda}_s$ is the linear transformation Λ_s of Eq. (3.4), appropriately modified for the axis reversal. It is clear that $\tilde{\Lambda}_s$ leaves the y and z coordinates unchanged, that is, $y = \tilde{y}'$ and $z = \tilde{z}'$. We therefore reduce the problem to two dimensions and consider $\tilde{\Lambda}_s$ as a function from $\tilde{K}'$ to K, with $\tilde{\Lambda}_s(\tilde{t}', \tilde{x}') = (t, x)$. Next, we invoke the Principle of Relativity, which states that all inertial frames are equivalent. This implies that the spacetime transformations between two inertial frames can depend only on the relative velocity between them. Since the velocity coordinates of $\tilde{K}'$ in K are equal to the velocity coordinates of K in $\tilde{K}'$, the inverse transformation $\tilde{\Lambda}_s^{-1} : K \to \tilde{K}'$ is the same as $\tilde{\Lambda}_s : \tilde{K}' \to K$. In other words,

$$\tilde{\Lambda}_s = \tilde{\Lambda}_s^{-1} \quad \text{or} \quad \tilde{\Lambda}_s^2 = I. \tag{3.6}$$

This means that $\tilde{\Lambda}_s$ is a *symmetry*.

Since $\tilde{\Lambda}_s$ is linear, we may represent it in the form

$$\begin{Bmatrix} t = a\tilde{t}' + b\tilde{x}' \\ x = c\tilde{t}' + d\tilde{x}' \end{Bmatrix}. \tag{3.7}$$

To determine the values of the coefficients a, b, c and d, we first consider the motion of $O' = \begin{pmatrix} \tilde{t}' \\ 0 \end{pmatrix}$ in K. From (3.7), we have $t = a\tilde{t}'$ and $x = c\tilde{t}'$. We assume that the time t in K and the time $\tilde{t}'$ in $\tilde{K}'$ flow in the same direction. In other words, if t increases, so does $\tilde{t}'$. This implies that

$$a > 0. \tag{3.8}$$

Moreover, the velocity v of O' in K is $v = x/t = c/a$, so from (3.7) and $c = av$, the matrix of $\tilde{\Lambda}_s$ is

$$\tilde{\Lambda}_s = \begin{pmatrix} a & b \\ av & d \end{pmatrix}. \tag{3.9}$$

Thus, from the second equation of (3.6), we have

(I) $a^2 + abv = 1$
(II) $b(a + d) = 0$
(III) $av(a + d) = 0$
(IV) $abv + d^2 = 1$

Dividing **(I)** by a^2, we get $1 + (b/a)v = 1/a^2$. Let $\kappa = b/a$. Then $a^2 = 1/(1 + \kappa v)$, or

$$a = \frac{1}{\sqrt{1 + \kappa v}}. \tag{3.10}$$

We use only the positive square root in light of (3.8).

Since $v \neq 0$, **(III)** implies that $d = -a$. Thus, $\tilde{\Lambda}_s = \frac{1}{\sqrt{1+\kappa v}} \begin{pmatrix} 1 & \kappa \\ v & -1 \end{pmatrix}$. By applying space reversal and a rotation, we re-reverse the x' axis, and in standard configuration, we have

$$\Lambda_s = \frac{1}{\sqrt{1 + \kappa v}} \begin{pmatrix} 1 & -\kappa \\ v & 1 \end{pmatrix}. \tag{3.11}$$

At this point, the derivation of the spacetime transformations splits into two cases: $\kappa = 0$ and $\kappa \neq 0$.

3.2.1 *The Galilean Transformations*

Assume now that $\kappa = 0$. Then $a = 1$, and the matrix Λ_s^G (G for Galilean) is

$$\Lambda_s^G = \begin{pmatrix} 1 & 0 \\ v & 1 \end{pmatrix}. \tag{3.12}$$

The *Galilean transformations* are, therefore,

$$\begin{aligned} &t = t' \\ &x = vt' + x' \ , \quad y = y' \ , \quad z = z'. \end{aligned} \tag{3.13}$$

These are the transformations associated with Newtonian dynamics. Note that time is absolute. This assumes that it is possible to synchronize all clocks in all inertial frames. Equivalently, there exists a universal clock which displays the time of all events.

To visualize the Galilean transformations, refer to Fig. 3.6. Draw perpendicular axes t, x for K. Since K and K' are in standard configuration, and the velocity of K' in K is $\mathbf{v} = (v, 0, 0)$, it follows that in K, the worldline of the origin O' of K' is (t, vt), or $x = vt$. On the other hand, the worldline of O' in K is also the t' axis in K'. Thus, the Eq. of the t' axis is $x = vt$. The x' axis is found by substituting $t' = 0$ into (3.13), which implies that $t = 0$. Thus, the two frames share a common x, x'

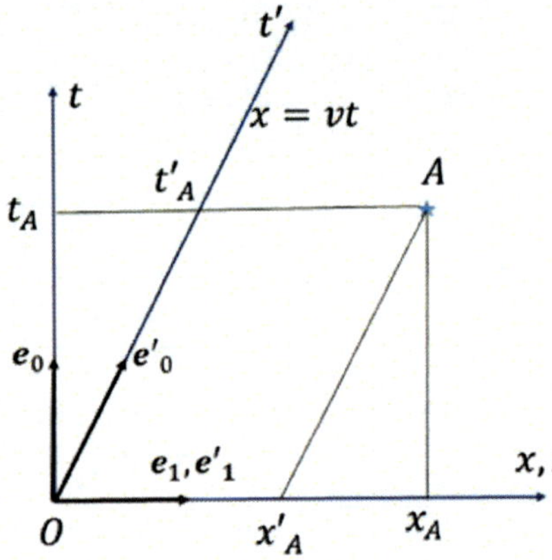

Fig. 3.6 The Galilean transformations (3.13). In K, the basis vector $\mathbf{e}_0$ has coordinates $(1, 0)$, and $\mathbf{e}_0'$ has coordinates $(1, v)$. The basis vector $\mathbf{e}_1'$ has coordinates $(0, 1)$. Thus, the matrix of the Galilean transformation is Λ_s^G of (3.12)

axis. Visually, one locates the coordinates of an event in an inertial frame by drawing lines from the event parallel to the frame's axes. In order that $t = t'$, one unit on the t' axis must be equal to $\sqrt{1 + v^2}$ units on the t axis.

Equation (3.13) are the connection between the spacetime coordinates of K and K', where $\mathbf{v} = (v, 0, 0)$ is the velocity of K' in K. Next, we check how velocities $\mathbf{u} = (u_x, u_y, u_z) = \left(\frac{dx}{dt}, \frac{dy}{dt}, \frac{dz}{dt}\right)$ and accelerations $\mathbf{a} = \frac{d\mathbf{u}}{dt}$ transform. Suppose an object moves with velocity $\mathbf{u}'$ in K'. Its worldline in K' is thus $(t', \mathbf{x}'(0) + \mathbf{u}'t')$. Using the Galilean transformations (3.13), its worldline in K is $(t, \mathbf{x}'(0) + (\mathbf{v} + \mathbf{u}')t)$. Hence, the velocity $\mathbf{u}$ of this object in K is

$$\mathbf{u} = \mathbf{v} + \mathbf{u}'. \tag{3.14}$$

Formula (3.14) is usually called the *velocity addition formula*, for obvious reasons. However, it is actually a *composition* of velocities, since $\mathbf{u}'$ is a velocity in K', while $\mathbf{v}$ is the relative velocity of K' in K.

Next, consider motion with varying velocity $\mathbf{u}'(t)$ in K'. Differentiating (3.14) by $t = t'$ and recalling that $\mathbf{v}$ is a constant, we see that observers in K and K' measure the same acceleration:

$$\mathbf{a} = \frac{d\mathbf{u}(t)}{dt} = \frac{d(\mathbf{v} + \mathbf{u}'(t))}{dt'} = \frac{d\mathbf{u}'(t)}{dt'} = \mathbf{a}'.$$

Thus, it is obvious that Newton's Second Law $\mathbf{F} = m\mathbf{a}$ will be the same in K and K' since all components are the same for both observers.

Newtonian dynamics, equipped with the Galilean transformations, is a very successful theory. In fact, it is used today in most technologies. Nevertheless, the Galilean transformations have been shown experimentally to be incorrect. The Michelson-Morley experiment [76] showed that the speed of light c in vacuum is independent of the velocity of the emitter of light. This means that light emitted in the positive

x direction by an emitter at rest at the origin of K' has velocity $(c, 0, 0)$ in both K and K'. Thus, the velocity addition formula (3.14) implies that $v + c = c$, which is impossible for $v \neq 0$.

An experiment in 1851 by Armand Hippolyte Louis Fizeau [32] measured the speed of light in moving water. The speed of light in resting water is $\frac{c}{n}$, where $n \approx 1.3$ is the refractive index of water. The velocity addition formula (3.14) predicts that when light travels with the flow of water, the speed of light in the lab should be $v + \frac{c}{n}$, where v is the velocity of the water in the lab. Fizeau, however, observed the speed of light in the lab to be $\frac{c}{n} + v\left(1 - \frac{1}{n^2}\right)$.

Incidentally, there are additional difficulties with Newtonian dynamics and the Galilean transformations. According to Newton's Second Law $\mathbf{F} = m\mathbf{a}$, the force does not depend on an object's velocity. But in the late nineteenth century, it was discovered that the magnetic force on a charge depends on the charge's velocity. Moreover, Maxwell's equations (1862) are not invariant under the Galilean transformations.

We conclude, therefore, that $\kappa \neq 0$ (see formula (3.11)) and proceed in the next section to derive the Lorentz transformations.

3.2.2 *The Lorentz Transformations*

Consider now the case $\kappa \neq 0$. Since, by the Principle of Relativity, the laws of physics are the same in all inertial frames, we seek a parameter of evolution which will be the same for all inertial observers. In Newtonian dynamics, this parameter is the time t, which is the same for all observers. But from (3.11), if $\kappa \neq 0$, time is not an invariant quantity.

We look for a parameter to express the separation, or *interval*, between two events. This interval is expressed by a vector $\Delta s = (\Delta t, \Delta x, \Delta y, \Delta z)$. It is preferable that all four spacetime coordinates have the same dimension, and so we introduce a constant μ with units of velocity and replace the zero component of the displacement with $\mu \Delta t$. We shall prove the following claim.

Claim

(a) The transformations Λ (defined by (3.4)) and Λ_s (defined by (3.11)), with $\kappa \neq 0$, preserve the square interval Δs^2 between two events defined by

$$\Delta s^2 = \mu^2 \Delta t^2 - \Delta x^2 - \Delta y^2 - \Delta z^2 \tag{3.15}$$

if and only if

$$\kappa = -\frac{v}{\mu^2}, \quad \text{or} \quad \mu = \sqrt{\frac{-v}{\kappa}}. \tag{3.16}$$

(b) If $\kappa = -\frac{v}{\mu^2}$ and an object moves with constant speed μ as measured in K', its speed in K is also μ.

We require that the interval Δs^2 in (3.15) be invariant under all transformations of the form (3.4). Since separations are invariant under translations (Eq. (3.2)), it is enough to check that Δs^2 is invariant under rotations and transformations Λ_s (3.11) between frames in standard configuration. Rotations do not change the time, and they preserve spatial distances. Therefore, the interval Δs^2 is invariant under rotations. It remains to check the transformations Λ_s of the form (3.11).

Now

$$\begin{pmatrix} \Delta t \\ \Delta x \end{pmatrix} = a \begin{pmatrix} 1 & -\kappa \\ v & 1 \end{pmatrix} \begin{pmatrix} \Delta t' \\ \Delta x' \end{pmatrix},$$

or

$$\begin{aligned} \Delta t &= a\Delta t' - a\kappa\Delta x' \\ \Delta x &= av\Delta t' + a\Delta x' \end{aligned}.$$

Substituting these into (3.15) and suppressing the y and z directions, we have

$$\Delta s^2 = \mu^2 \Delta t^2 - \Delta x^2$$

$$= \mu^2 \left(a^2(\Delta t')^2 - 2a^2\kappa\Delta t'\Delta x' + a^2\kappa^2(\Delta x')^2\right) - \left(a^2v^2(\Delta t')^2 + 2a^2v\Delta t'\Delta x' + a^2(\Delta x')^2\right)$$

$$= (\mu^2a^2 - a^2v^2)(\Delta t')^2 - (2\mu^2a^2\kappa + 2a^2v)\Delta t'\Delta x' + (\mu^2a^2\kappa^2 - a^2)(\Delta x')^2.$$

Thus, in order that $(\Delta s')^2 = \Delta s^2$, the following identities must hold:

(1) $\mu^2a^2 - a^2v^2 = \mu^2$
(2) $2\mu^2a^2\kappa + 2a^2v = 0$, or $\mu^2\kappa + v = 0$
(3) $\mu^2a^2\kappa^2 - a^2 = -1$.

Thus, if $(\Delta s')^2 = \Delta s^2$, then (2) holds, which immediately implies (3.16). Conversely, suppose (3.16) holds. Then

$$a^2 = \frac{1}{1+\kappa v} = \frac{1}{1 - \frac{v^2}{\mu^2}}. \tag{3.17}$$

It is then easy to verify that (1)–(3) hold with the values of a^2 and κ from (3.17) and (3.16). Thus, $(\Delta s')^2 = \Delta s^2$. This proves (a).

For (b), consider uniform motion in K', with speed μ. The motion is automatically uniform in K, by the Principle of Relativity. We show that in K, the speed of this uniform motion is also μ. Pick any two events on the worldline of this motion. By (3.15), the invariant interval between these events is $(\Delta s')^2 = 0$. By invariance of the interval, we have $\Delta s^2 = 0$ in K, implying that the object has speed μ in K as well. This proves (b) and the claim.

Part (b) of the claim means that if we want the transformation Λ to preserve the interval (3.15), then the *speed* μ should be the same in K and K'. We know from the Michelson-Morley [76] experiment that the speed of light c in vacuum has this property. Therefore, we may assume that $\mu = c$. In the next section, we show that this choice is unique.

From (3.11) and (3.16), with $\mu = c$, the transformation $\Lambda_s : K' \to K$ between frames in standard configuration is

$$\begin{aligned} t &= \gamma\left(t' + \tfrac{vx'}{c^2}\right) \\ x &= \gamma(vt' + x') \\ y &= y' \\ z &= z' \end{aligned}, \tag{3.18}$$

where

$$\gamma = \gamma(v) = \frac{1}{\sqrt{1 - \frac{v^2}{c^2}}}. \tag{3.19}$$

The transformation (3.18) can be written in matrix form as

$$\begin{pmatrix} t \\ x \\ y \\ z \end{pmatrix} = \begin{pmatrix} \gamma & \gamma v & 0 & 0 \\ \gamma v & \gamma & 0 & 0 \\ 0 & 0 & 1 & 0 \\ 0 & 0 & 0 & 1 \end{pmatrix} \begin{pmatrix} t' \\ x' \\ y' \\ z' \end{pmatrix} \tag{3.20}$$

and is a *Lorentz boost*, that is, a Lorentz transformation between inertial frames which have no relative rotation. The collection of all Lorentz boosts, together with the orthogonal group $O(3)$ of spatial rotations in 3D Euclidean space, form the *Lorentz group*. The *Poincaré group* contains the Lorentz group as a linear subgroup, as well as the spacetime translations. The Poincaré transformations preserve the interval

$$\Delta s^2 = c^2\Delta\tau^2 = c^2\Delta t^2 - \Delta x^2 - \Delta y^2 - \Delta z^2. \tag{3.21}$$

Note that when the relative velocity v between the frames is small compared to the speed of light, we can ignore terms of order v/c, and the Lorentz transformations reduce to the Galilean transformations (3.13). Thus, for small velocities, Newtonian dynamics is a perfectly valid theory.

To visualize these transformations, refer to Fig. 3.7. In figures presenting relativistic effects, for convenience we use units of time in which the speed of light $c = 1$. Draw perpendicular axes t, x for K. Since the velocity of K' in K is $\mathbf{v} = (v, 0, 0)$, it follows that the equation of the t' axis is $x = vt$, as it was for the Galilean transformations (Fig. 3.6). The x' axis is found by substituting $t' = 0$ into (3.18), which implies that $t = vx$, which is different from the Galilean transformations. Note that the t' and the x' axes are symmetric with respect the line LC $t = x$, which represents

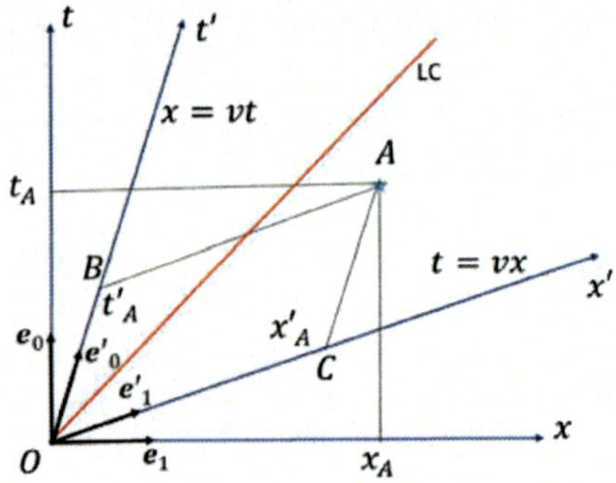

Fig. 3.7 The Lorentz transformations (3.18). The basis vector $\mathbf{e}'_0$ has coordinates (t, vt) in K and coordinates $(1, 0)$ in K'. Since the Lorentz transformations preserve the interval (3.21), we have $t^2 - v^2t^2 = 1$, and $t = \gamma$. Thus, $\mathbf{e}'_0$ has coordinates $(\gamma, \gamma v)$ in K. Similarly, the basis vector $\mathbf{e}'_1$ has coordinates (vx, x) in K and coordinates $(0, 1)$ in K'. By preservation of the interval, $x = \gamma$, and $\mathbf{e}'_1$ has coordinates $(\gamma v, \gamma)$ in K. Thus, the matrix of the Lorentz transformation is the 4×4 matrix on the right-hand side of (3.20). Notice also that the Lorentz transformation can be derived from $OA = OB + OC = t'_A\mathbf{e}'_0 + x'_A\mathbf{e}'_1$

the light cone. Visually, one locates the coordinates of an event in an inertial frame by drawing lines from the event parallel to the frame's axes.

In the next section, which may be skipped without loss of continuity, we derive Einstein velocity addition and explore some of its applications.

3.3 Einstein Velocity Addition and Applications

3.3.1 Velocity Addition

As in the case of the Galilean transformations, *Einstein velocity addition* is actually a *composition* of velocities and is defined in the following way. Let K and K' be in standard configuration, where K' has velocity $\mathbf{v} = (v, 0, 0)$ in K. Suppose an object has velocity $\mathbf{u}$ in K'. Then the velocity of the object in K is denoted by $\mathbf{v} \oplus \mathbf{u}$.

To derive a formula for $\mathbf{v} \oplus \mathbf{u}$, consider an object with velocity $\mathbf{u} = (u_1, u_2, u_3)$ in K'. Without loss of generality, the object passes through the origin O' of K' at time $t' = 0$. Hence, the worldline of the object in K' is $(t', u_1t', u_2t', u_3t')$. Using the Lorentz transformation (3.18), we have

$$\begin{aligned} t &= \gamma t' + \gamma \tfrac{v}{c^2} u_1 t' \\ x &= \gamma v t' + \gamma u_1 t' \\ y &= u_2 t' \\ z &= u_3 t' \end{aligned}, \tag{3.22}$$

with γ defined by (3.19). Thus, the velocity of the object in K is

$$\frac{(x, y, z)}{t} = \frac{(\gamma(v + u_1), u_2, u_3)}{\gamma + \gamma\frac{v}{c^2}u_1},$$

implying that

$$\mathbf{v} \oplus \mathbf{u} = \frac{(v + u_1, \gamma^{-1}u_2, \gamma^{-1}u_3)}{1 + \frac{vu_1}{c^2}}. \tag{3.23}$$

For arbitrary $\mathbf{v}$, $\mathbf{u}$, decompose $\mathbf{u}$ as $\mathbf{u} = \mathbf{u}_\| + \mathbf{u}_\perp$, where $\mathbf{u}_\|$ is the projection of $\mathbf{u}$ onto $\mathbf{v}$ and $\mathbf{u}_\perp = \mathbf{u} - \mathbf{u}_\|$. Then formula (3.23) generalizes to

$$\mathbf{v} \oplus \mathbf{u} = \frac{\mathbf{v} + \mathbf{u}_\| + \gamma^{-1}\mathbf{u}_\perp}{1 + \frac{\mathbf{v} \circ \mathbf{u}}{c^2}}, \tag{3.24}$$

where $\mathbf{v} \circ \mathbf{u}$ is the Euclidean scalar product on $\mathbb{R}^3$, defined below in formula (3.55). If $\mathbf{v}$ and $\mathbf{u}$ are parallel, then

$$\mathbf{v} \oplus \mathbf{u} = \frac{\mathbf{v} + \mathbf{u}}{1 + \frac{vu}{c^2}}. \tag{3.25}$$

Formulas (3.24) and (3.25) show that Einstein velocity addition is commutative only for parallel velocities.

We show now that the speed of light c in vacuum is the unique speed which is the same in all inertial frames. Suppose that the velocity between the frames is $\mathbf{v} \neq 0$ and that the velocity of an object $\mathbf{u}$ in K' is parallel to $\mathbf{v}$. From (3.25), $\mathbf{v} \oplus \mathbf{u}$ is also parallel to $\mathbf{v}$. Thus, if the speed of the object is u also in K, then $\mathbf{v} \oplus \mathbf{u} = \mathbf{u}$, or $\frac{v+u}{1+\frac{vu}{c^2}} = u$. Multiplying both sides by the denominator leads to $v = v\frac{u^2}{c^2}$, implying that $u = \pm c$. Thus, only the speed of light is preserved by the Lorentz transformations. This also implies that the speed of light is independent of the speed of the emitter, as follows. Let the emitter of the light move with velocity $\mathbf{v}$ in our lab frame. The speed of light in the frame attached to the emitter is $|\mathbf{u}| = c$. From the definition of the velocity addition, the velocity of the light in the lab frame is $\mathbf{v} \oplus \mathbf{u}$, with speed c.

Note that velocity addition (3.24) maps a subluminal (less than c) velocity to a subluminal velocity. Subluminal velocities are appropriate for massive particles. There are also particles, such as photons, with speed c in every inertial frame. The theory presented in this book does not preclude the existence of superluminal velocities, nor does it predict them. Therefore, for the sake of simplicity, we define the *ball D_u of admissible* 3D *velocities* as

$$D_u = \{\mathbf{u} \in \mathbb{R}^3 : |\mathbf{u}| \leq c\}. \tag{3.26}$$

For a fixed $\mathbf{v} \in D_u$, we define a function $\varphi_\mathbf{v} : D_u \to D_u$ by

$$\varphi_\mathbf{v}(\mathbf{u}) = \mathbf{v} \oplus \mathbf{u}.$$

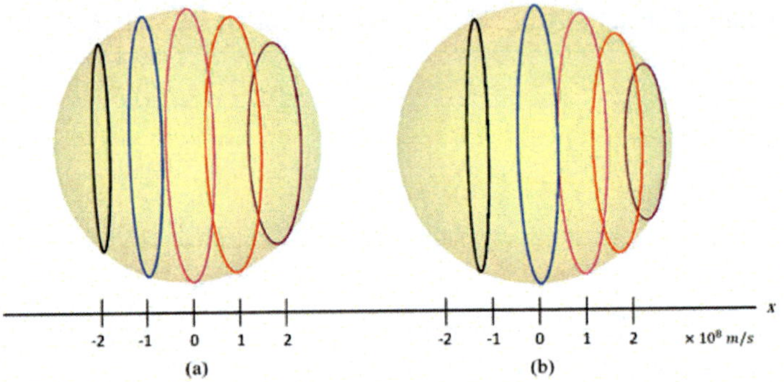

Fig. 3.8 **a** A set of 5 uniformly spread discs Δ_j obtained by intersecting the three-dimensional velocity ball D_u of radius $c = 3 \cdot 10^8$m/s with u_2, u_3 planes at $u_1 = 0, \pm 10^8, \pm 2 \cdot 10^8$m/s. **b** The images of these Δ_j under the map $\varphi_{\mathbf{v}}(\mathbf{u}) = \mathbf{v} \oplus \mathbf{u}$, with $\mathbf{v} = (10^8, 0, 0)$m/s. Note that $\varphi_{\mathbf{v}}(\Delta_j)$ is also a disc in D_u, perpendicular to $\mathbf{v}$ and moved in the direction of $\mathbf{v}$. On each disc Δ_j, the map $\varphi_{\mathbf{v}}$ acts as multiplication by a constant in the $\mathbf{u}_\perp$ component. A similar figure appears in [33]

To understand the function $\varphi_{\mathbf{v}}$, decompose $\mathbf{u}$, as above, as

$$\mathbf{u} = \mathbf{u}_\| + \mathbf{u}_\perp.$$

For simplicity, let $\mathbf{v} = (v, 0, 0)$ and $\mathbf{u} = (u_1, u_2, u_2)$. Then, (3.24) and (3.25) imply that

$$\varphi_{\mathbf{v}}(\mathbf{u}) = \left(\frac{v + u_1}{1 + v u_1/c^2}, 0, 0 \right) + \frac{\gamma^{-1}}{1 + v u_1/c^2}(0, u_2, u_3). \tag{3.27}$$

Consider the intersection of the plane $u_1 = a$ with D_u, which is a disc Δ_a. From the previous formula, the x component of every element of $\varphi_{\mathbf{v}}(\Delta_a)$ is $a' = \frac{v+a}{1+va/c^2}$. Thus, $\varphi_{\mathbf{v}}(\Delta_a)$ is also a disc perpendicular to the x axis. See Fig. 3.8.

We revisit now the Fizeau experiment, in which the measured speed of light in moving water differed from that predicted by the Galilean transformations (see end of Sect. 3.2.1). As mentioned above, the speed of light in resting water is $\frac{c}{n}$, where $n \approx 1.3$ is the refractive index of water. We choose the axes so that the velocity of the water in our lab frame K is $\mathbf{v} = (v, 0, 0)$. Let K' be a frame moving with velocity $\mathbf{v}$. In K', the water is at rest, and, thus, the light has velocity $\mathbf{u}$, which we assume to be parallel to $\mathbf{v}$, as it was in the original experiment, with $|\mathbf{u}| = \frac{c}{n}$. By the Einstein velocity addition formula (3.25), the speed of light in the lab frame is $\mathbf{v} \oplus \mathbf{u}$, and its magnitude is

$$\frac{v + c/n}{1 + v/nc}. \tag{3.28}$$

Assuming that $v \ll c$, we expand this expression in a power series and discard terms of order one or higher in v/c. This approximation yields the speed $v + \frac{c}{n} - \frac{v}{n^2} = \frac{c}{n} + v\left(1 - \frac{1}{n^2}\right)$, which is exactly the speed observed in the experiment. Thus, Fizeau's experiment confirms that the true transformations between inertial frames are the Lorentz transformations.

3.3.2 *Fiber Optic Gyroscopes and the Sagnac Effect*

In 1913, Georges Sagnac showed that if a beam of light is split and sent in opposite directions around a closed path on a rotating platform, the two beams will exhibit interference when recombined. This is called the *Sagnac effect*. It provides a method of detecting and measuring rotations.

In a fiber optic gyroscope (FOG), a beam of coherent light is split, and the two beams are made to follow the same circular fiber optics path, but in opposite directions. The beams exhibit interference at the exit point. This interference depends on the phase shift between the beams at the exit point. Following the approach of [73], we will find the dependence of the phase shift on the angular velocity of the apparatus.

Refer to Fig. 3.9. Denote the entrance point by A and the exit point by B, where A and B are spatially separated only vertically by a very small distance. The interference manifests itself in the phase difference between the two beams at B. Thus, we need to follow the phase propagation of the beam in the fiber. Let v_p be the phase velocity of the beam in a fiber at rest in some inertial system. It is known that v_p is independent of the bending of the fiber. Thus, we may assume that the fiber lies on the x axis and that the velocity of the fiber is in the x direction. For rotation with given frequency Ω, this velocity is $v_r = R\Omega$, where R is the radius of the circular fiber.

Calculate first the time t^+ it will take the given phase Φ_0 to travel from A to B, for the beam in the direction of v_r. This time may be obtained by dividing the distance

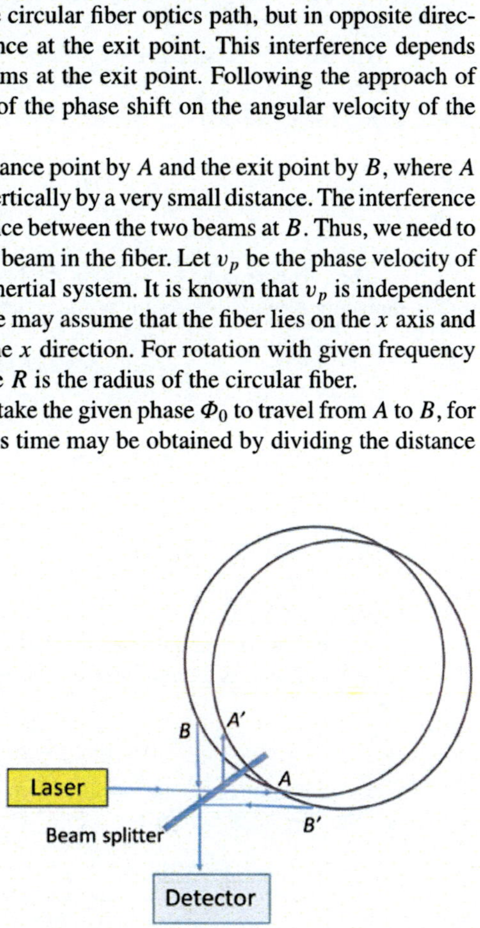

Fig. 3.9 Schematic of a fiber optic gyroscope (FOG). A coherent beam of light is emitted by the laser. The beam is split and circles the FOG in opposite directions. Any rotation with respect to an inertial frame will cause a phase shift when the beams rejoin. The rotational velocity is then calculated from the phase shift, the frequency of the light, and the area of the FOG

between these points in our inertial lab frame K by the relative velocity between Φ_0 and the point B in K.

By the Einstein velocity addition formula (3.25), the velocity v^+ of Φ_0 in K is

$$v^+ = v_r \oplus v_p = \frac{v_r + v_p}{1 + v_r v_p/c^2}.$$

The relative velocity between two points in an inertial system is the usual vector difference between these velocities. Thus, the relative velocity between Φ_0 and the point B in K is

$$v^+ - v_r = \frac{v_r + v_p}{1 + v_r v_p/c^2} - v_r = \frac{v_p(1 - v_r^2/c^2)}{1 + v_r v_p/c^2}.$$

The distance between A and B in K is the distance in rest frame $d_0 = 2\pi R$ contracted by $d = d_0\sqrt{1 - v_r^2/c^2}$ of the fiber motion. Thus, this distance is $2\pi R\sqrt{1 - v_r^2/c^2}$ and

$$t^+ = \frac{2\pi R/\gamma}{v^+ - v_r} = \frac{2\pi R(1 + v_r v_p/c^2)}{v_p\sqrt{1 - v_r^2/c^2}}. \tag{3.29}$$

The corresponding formula for the beam in the opposite direction is obtained by replacing v_r with $-v_r$ in all formulas, leading to

$$t^- = \frac{2\pi R(1 - v_r v_p/c^2)}{v_p\sqrt{1 - v_r^2/c^2}}. \tag{3.30}$$

Thus, the difference in the times is

$$\Delta t = t^+ - t^- = \frac{2\pi R 2 v_r v_p/c^2}{v_p\sqrt{1 - v_r^2/c^2}} = \frac{4\pi R^2 \Omega}{c^2\sqrt{1 - v_r^2/c^2}}, \tag{3.31}$$

which is independent of v_p. This fact has also been observed experimentally.

The phase shift should be calculated in the frame K' comoving with the fiber, since the detectors are mounted on the apparatus. Thus, the time difference in K' is

$$\Delta t' = \sqrt{1 - v_r^2/c^2}\,\Delta t = \frac{4\pi R^2 \Omega}{c^2},$$

and the Sagnac phase shift is

$$\Phi_S = \omega \Delta t' = \omega \frac{4\pi R^2 \Omega}{c^2}, \tag{3.32}$$

where ω is the frequency of the beam in K. Thus, the rotation with respect to K is

$$\Omega = \frac{\Phi_S c^2}{4\pi R^2 \omega}. \tag{3.33}$$

This agrees with the known formula for the Sagnac effect.

3.4 Minkowski Space

In this section, we give a formal definition of *Minkowski space*, also called *flat spacetime*. This is spacetime without sources—no gravitational sources, no electromagnetic sources, and no media. Just empty spacetime.

The geometry of flat spacetime is governed by the interval

$$\Delta s^2 = c^2 \Delta t^2 - \Delta x^2 - \Delta y^2 - \Delta z^2 \tag{3.34}$$

between two events. We have seen that the Poincaré transformations preserve this interval. Thus, the interval between two events is a *Lorentz-invariant scalar*, that is, a real-valued function that has the same value in every inertial frame.

To write the interval Δs^2 in tensor form, we define $x^0 = ct, x^1 = x, x^2 = y, x^3 = z$ and

$$\eta_{\mu\nu} = \left\{ \begin{array}{ll} 1, & \text{if } \mu = \nu = 0 \\ -1, & \text{if } \mu = \nu \in \{1, 2, 3\} \\ 0, & \text{otherwise} \end{array} \right\}, \tag{3.35}$$

where μ and ν range over $0, 1, 2, 3$. In this notation, one has

$$\Delta s^2 = \eta_{\mu\nu} \Delta x^\mu \Delta x^\nu, \tag{3.36}$$

where we have adopted *Einstein's summation convention* in which there is an implied summation over all repeated upper and lower indices. Thus, in (3.36), the expression is summed over μ, since μ appears once as an upper index and once as a lower index. For the same reason, the expression is also summed over ν. Note that (3.35) is usually written more concisely as $\eta_{\mu\nu} = \mathrm{diag}(1, -1, -1, -1)$.

The *proper time interval* $\Delta\tau$ between two events is defined by $\Delta\tau = \sqrt{\Delta s^2/c^2}$, where $\Delta s^2 \geq 0$ is defined by (3.36). Since $\Delta s^2 \geq 0$, there is an inertial frame in which, for these two events, $\Delta x = \Delta y = \Delta z = 0$. Thus, in this frame, $\Delta s^2 = c^2 \Delta t^2$, which implies that $\Delta\tau = \Delta t$. Thus, the proper time interval between two events is the time interval measured by a inertial clock traveling between the two events. Since the interval and the speed of light are Lorentz-invariant scalars, so is the proper time interval.

In infinitesimal form, the Lorentz-invariant interval between two infinitesimally close spacetime points (ct, x, y, z) and $(c(t + dt), x + dx, y + dy, z + dz)$ is

$$ds^2 = c^2 dt^2 - dx^2 - dy^2 - dz^2 = \eta_{\mu\nu} dx^\mu dx^\nu \tag{3.37}$$

and is known as the *Minkowski metric* or the *metric of flat spacetime*. Dividing the proper time interval by Δs, we define the *proper time* τ by $cd\tau = ds$. Clearly, the proper time is a Lorentz-invariant scalar. We very often use proper time as the parameter on an object's worldline. This is a natural choice of parameter, since, in this case, the proper time is the time displayed on a clock comoving with the object. However, it is more convenient for the parameter to have dimensions of length. Therefore, we use $c\tau$ instead of τ as the parameter. Moreover, we rename $c\tau$ as τ. This is equivalent to using units in which $c = 1$, a common practice in relativity. To reiterate, in this book, *the parameter* τ *has dimensions of length*. If one wants to return to dimensions of time, he must remember to divide by c. Thus, the proper time parameter τ is defined by

$$d\tau^2 = \eta_{\mu\nu} dx^\mu dx^\nu = c^2 dt^2 \left(1 - \frac{v^2}{c^2}\right), \tag{3.38}$$

where $\mathbf{v} = (dx/dt, dy/dt, dz/dt)$ and $v = |\mathbf{v}|$. From this formula, one sees that

$$\frac{d\tau}{dt} = c\gamma^{-1}, \tag{3.39}$$

where γ is defined by (3.19). From (3.38), it follows immediately that $d\tau^2 = 0$ if and only if $v = c$. Incidentally, it should come as no surprise that the time t in some (any) inertial frame is *not* a Lorentz-invariant scalar and is therefore not a suitable choice for the parameter of the worldline of a moving object.

Consider an object with worldline $x(\tau)$. Recall from (3.3) that the object's *four-velocity* with respect to τ is

$$u(\tau) = \frac{dx}{d\tau}. \tag{3.40}$$

From (3.39), we have $dx^0/d\tau = c\,dt/d\tau = \gamma$, and $dx^i/d\tau = dx^i/dt \cdot dt/d\tau = \gamma v^i/c$. Hence, the four-velocity is

$$u(\tau) = \gamma\left(1, \frac{\mathbf{v}}{c}\right) = \gamma\,(1, \boldsymbol{\beta}), \tag{3.41}$$

where $\mathbf{v}$ is the 3D velocity and $\boldsymbol{\beta} = \mathbf{v}/c$. Note that the four-velocity is unit free. An object's *four-acceleration with respect to* τ is the four-vector

$$a(\tau) = \frac{du}{d\tau}. \tag{3.42}$$

Note that the four-acceleration has dimensions of 1/length. For dimensions of acceleration, one must multiply the four-acceleration by c^2.

Another example of a Lorentz-invariant scalar is provided by the Minkowski inner product, defined as follows. Let x, y be two four-vectors. The *Minkowski inner product* $x \cdot y$ of $x = x^\mu$ and $y = y^\nu$ is defined by

$$x \cdot y = x^0 y^0 - x^1 y^1 - x^2 y^2 - x^3 y^3 = \eta_{\mu\nu} x^\mu y^\nu. \tag{3.43}$$

The Minkowski inner product is *bilinear*, that is, it is linear in both the first and the second variable. The *squared norm* of a four-vector is $(x)^2 = x \cdot x$. Note that the four-velocity (3.41) has squared norm equal to 1:

$$(u)^2 = \gamma^2 - \gamma^2 \frac{v^2}{c^2} = 1.$$

Differentiating $(u)^2 = 1$, we obtain $a \cdot u = 0$, implying that the four-velocity and the four-acceleration are orthogonal. More generally, for any parameter σ, the four-velocity $dx/d\sigma$ with respect to σ has positive squared norm. Indeed, since

$$\frac{dx}{d\sigma} = \frac{dx}{dt}\frac{dt}{d\sigma} = \frac{dt}{d\sigma}(c, \mathbf{v}),$$

$dx/d\sigma$ has squared norm

$$\left(\frac{dt}{d\sigma}\right)^2 (c^2 - v^2) > 0.$$

By the preservation of the interval (3.36), the squared norm of a four-vector is a Lorentz-invariant scalar. From this, it follows that the inner product $x \cdot y$ is also a Lorentz-invariant scalar, since

$$x \cdot y = \frac{1}{2}\left((x + y)^2 - (x)^2 - (y)^2\right).$$

We now formally define *Minkowski space* or *flat spacetime* as the space $\mathbb{R}^4$ of spacetime positions, endowed with the Minkowski inner product (3.43) on the set $\tilde{M}$ of four-vectors.

For any spacetime point x, we define the *light cone* with vertex x as the set of spacetime points y such that $(y - x)^2 = 0$. The part of the light cone satisfying $y^0 > x^0$ is called the forward light cone, while the part satisfying $y^0 < x^0$ is called the backward light cone. A worldline lying on this cone represents motion with the speed of light, and the four-vector $y - x$ is called a *null vector*. Since the inner product is invariant, a Poincaré transformation takes a light cone into a light cone, with the vertex possibly shifted.

Since the four-velocity $dx/d\tau$ has squared norm 1, a Poincaré transformation maps a four-velocity to a four-velocity. See Figs. 3.10 and 3.11.

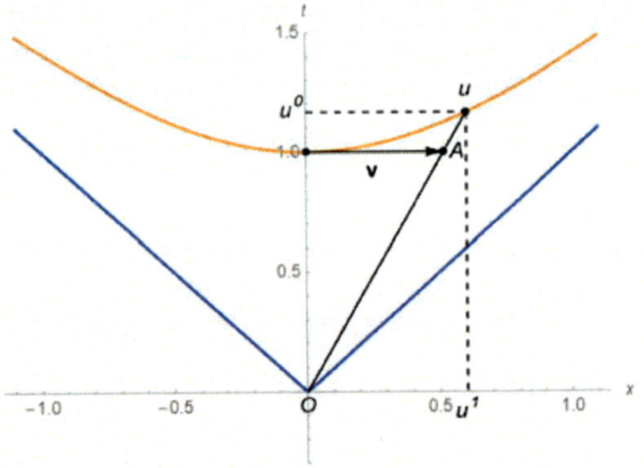

Fig. 3.10 (Blue) A two-dimensional section of the forward light cone $t^2 - x^2 = 0, t > 0$ with vertex at the origin. (Orange) A two-dimensional section of the hyperboloid $t^2 - x^2 = 1$ of four-velocities. The point A has coordinates $(t, x) = (1, \mathbf{v})$, and a multiple of the vector OA ends at the four-velocity $u = \gamma(1, \mathbf{v})$

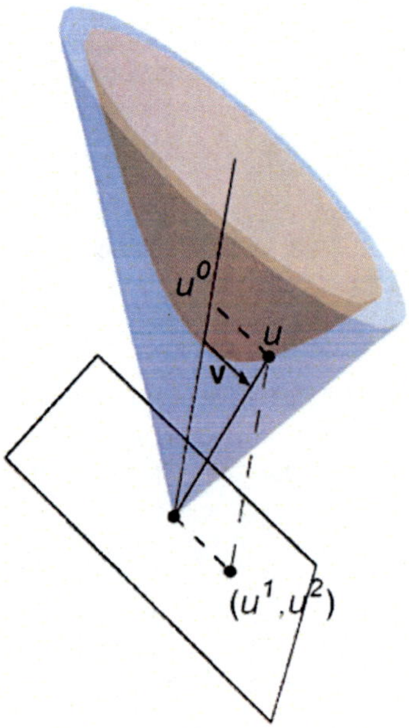

Fig. 3.11 (Light blue) A three-dimensional section of the forward light cone $t^2 - x^2 - y^2 = 0, t > 0$. (Lavender) A three-dimensional section of the hyperboloid $t^2 - x^2 - y^2 = 1$ of four-velocities. As in Fig. 3.10, the connection between u and $\mathbf{v}$ is shown

3.5 Four-Vectors, Four-Covectors, and Contraction

Four-vectors were defined in Sect. 3.1. Here we define the objects known as *four-covectors*. Using the Minkowski inner product, each four-vector canonically defines a four-covector. We also define the operation of *contraction*, which combines a four-covector and a four-vector to produce a scalar. We give several examples of contraction, from mathematics, physics, and life in general. We introduce the reader to the use of upper indices on the components of four-vectors and lower indices on the components of four-covectors. We show how to use $\eta_{\mu\nu}$ to lower indices.

Let V be a real or complex four-dimensional vector space. Let $B = \{\mathbf{e}_0, \mathbf{e}_1, \mathbf{e}_2, \mathbf{e}_3\}$ be the natural basis of V, where $\mathbf{e}_0 = (1, 0, 0, 0)$, $\mathbf{e}_1 = (0, 1, 0, 0)$, $\mathbf{e}_2 = (0, 0, 1, 0)$, $\mathbf{e}_3 = (0, 0, 0, 1)$. For every vector $x \in V$, there are unique scalars x^0, x^1, x^2, x^3 such that

$$x = \sum_{\mu=0}^{3} x^{\mu}\mathbf{e}_{\mu}. \tag{3.44}$$

The scalars x^{μ} are called the *coordinates* of x in the basis B. Note that these coordinates are written with an upper index, while the basis vectors $\mathbf{e}_{\mu}$ are written with a lower index. We do this in order to take advantage of *Einstein's summation convention*, in which one omits the summation sign and sums over any index which appears both above and below. Thus, (3.44) can be written as

$$x = x^{\mu}\mathbf{e}_{\mu}. \tag{3.45}$$

This provides for ease of notation.

A *four-covector* is a linear map $f : V \to \mathbb{R}$. A linear map is determined by its value on any set of basis four-vectors. Thus, given a linear map $f : V \to \mathbb{R}$, let $f_{\mu} = f(\mathbf{e}_{\mu})$, $\mu = 0, 1, 2, 3$. Here, we put a lower index on f to match the lower index on $\mathbf{e}$. The value of f on an arbitrary four-vector $x = x^{\mu}\mathbf{e}_{\mu}$ is then

$$f(x) = f\left(x^{\mu}\mathbf{e}_{\mu}\right) = x^{\mu} f(\mathbf{e}_{\mu}) = x^{\mu} f_{\mu}.$$

The middle equality uses the linearity of f. The expression $x^{\mu} f_{\mu}$ is called the *contraction* of the four-covector f and the four-vector x. Notice how the sum over μ is automatically implied since μ appears once as an upper index and once as a lower index.

There is a standard way of constructing a covector from a given vector. For example, fix a four-vector $a = (a^0, a^1, a^2, a^3)$ in Minkowski space. Define a function $a^* : \tilde{M} \to \mathbb{R}$ by

$$a^*(x) = a \cdot x = \eta_{\mu\nu}a^{\mu}x^{\nu}. \tag{3.46}$$

To compute the components a_{μ} of a^*, we apply a^* to the natural basis vectors $\mathbf{e}_{\mu}$. For $\mu = 0$, we have $\mathbf{e}_0 = (1, 0, 0, 0)$ and

$$a_0 = a^*(\mathbf{e}_0) = \eta_{\mu\nu}a^{\mu}(\mathbf{e}_0)^{\nu} = a^0,$$

and for $i = 1, 2, 3$,

$$a_i = a^*(\mathbf{e}_i) = \eta_{\mu\nu}a^{\mu}(\mathbf{e}_i)^{\nu} = \eta_{\mu i}a^{\mu} = -a^i.$$

Thus, $a^* = (a_0, a_1, a_2, a_3) = (a^0, -a^1, -a^2, -a^3)$ and

$$a^*(x) = a_{\nu}x^{\nu}. \tag{3.47}$$

Comparing this to (3.46), we see that

$$a_\nu = \eta_{\mu\nu} a^\mu. \tag{3.48}$$

Thus, η is used to lower indices, and, in general, we have

$$\eta_{\mu\nu} x^\mu y^\nu = x_\nu y^\nu = x^\mu y_\mu. \tag{3.49}$$

For a theory to satisfy the Principle of Relativity, one requires that the contraction $a_\nu x^\nu$ have the same value for all inertial observers. Let K and K' be two arbitrary inertial frames. Let Λ be the Lorentz transformation such that $x' = \Lambda x$. Λ maps the (unprimed) coordinates of $x \in K$ to their primed coordinates $x' \in K'$. To achieve Lorentz invariance of $a_\nu x^\nu$, we require that $(a^*)'$ be the four-covector in K' such that

$$(a^*)' x' = a^*(x).$$

Hence, for all $x \in K$, we have

$$(a^*)' \Lambda x = a^*(x).$$

This implies immediately that

$$(a^*)' \Lambda = a^* \quad \text{and} \quad (a^*)' = a^* \Lambda^{-1}. \tag{3.50}$$

This reveals an important rule: the components of a four-covector transform by the *inverse* of the Lorentz transformation of the components of a four-vector. This is another reason for using upper indices for the components of a vector and lower indices for the components of a covector. Since vectors and covectors transform differently, we ought to distinguish between them.

As an example of lowering indices, we prove the following differentiation rules for the Minkowski inner product:

$$\frac{d(x \cdot y)}{dx^k} = y_k \quad \text{and} \quad \frac{d(x)^2}{dx^k} = 2x_k. \tag{3.51}$$

To prove the first formula, we have $\frac{d(x \cdot y)}{dx^k} = \frac{d}{dx^k}(x^\mu y_\mu) = y_k$. For the second formula, first note that $\frac{dx_\mu}{dx^k} = \frac{d}{dx^k} \eta_{\mu\nu} x^\nu = \eta_{\mu k}$. Then, using the standard product rule,

$$\frac{d(x)^2}{dx^k} = \frac{d}{dx^k}(x_\mu x^\mu) = \frac{dx_\mu}{dx^k} x^\mu + x_\mu \frac{dx^\mu}{dx^k} = \eta_{\mu k} x^\mu + x_k = 2x_k.$$

We conclude this section with several examples of contraction.

Example 1 In the vegetable store.

A customer collects cucumbers, tomatoes, and red peppers. In order to calculate the total price, the proprietor weighs each vegetable, in pounds, say, and multiplies each pound amount by the price per pound of the corresponding vegetable. The scale shows that there are two pounds of cucumbers, three pounds of tomatoes, and four pounds of red peppers. Represent this as a weight vector

$$w = (w^1, w^2, w^3) = (2, 3, 4) \text{ pounds.} \tag{3.52}$$

If cucumbers cost forty cents per pound, tomatoes cost sixty cents per pound, and red peppers cost eighty cents per pound, represent this as a price per weight covector

$$p = (p_1, p_2, p_3) = (40, 60, 80) \text{ cents per pounds.} \tag{3.53}$$

The total price is then

$$w^i p_i = (2(40) + 3(60) + 4(80)) \text{ cents } = \$5.80.$$

The total price is thus calculated by contraction of a vector and a covector. Note that a change of basis will not change the total price. For example, suppose that we weigh the vegetables in kilograms instead of pounds. Using the relation 1 kg = 2.2 pounds and rounding, the weight vector is $w' = (0.91, 1.36, 1.82)$ kg, and the price per weight covector is $p' = (88, 132, 176)$ cents per kg. Ignoring rounding errors, the total price is

$$(w')^i (p)'_i = (0.91(88) + 1.36(132) + 1.82(176)) \text{ cents } = \$5.80,$$

as before. Note that with the above change of basis, the weight vector was *divided* by 2.2, while the price per weight covector was *multiplied* by 2.2. This displays the inverse relationship between the way that vectors and covectors transform.

Example 2 The gradient of a scalar function is a covector.

Let f be a scalar-valued function of several variables. We show that the gradient ∇f is a covector. For any x in the domain of f and small Δx, we have $f(x + \Delta x) - f(x) \approx \nabla f \cdot \Delta x$. Since the left-hand side of this formula is a scalar, and Δx is a vector, the gradient ∇f must be a covector. For example, the second formula of (3.51) shows that the kth component of $\nabla(x)^2$ is $2x_k$. That is, the components of $\nabla(x)^2$ have lower indices because $\nabla(x)^2$ is a covector. Similarly, a force $\mathbf{F}$ can be the gradient of a scalar potential. This implies that $\mathbf{F}$ is a covector.

Example 3 Work is the contraction of force and displacement.

In classical mechanics, the work W done by a force $\mathbf{F}$ along a path γ is

$$W = \int_\gamma \mathbf{F} \circ d\mathbf{r},$$

where $\circ$ is the Euclidean scalar product, defined explicitly in the next example. Here we contract the force covector $\mathbf{F} = (F_x, F_y, F_z)$ with the displacement vector $d\mathbf{r} = (dx, dy, dz)$:

$$\mathbf{F} \circ d\mathbf{r} = \begin{pmatrix} F_x & F_y & F_z \end{pmatrix} \begin{pmatrix} dx \\ dy \\ dz \end{pmatrix}.$$

This contraction represents the infinitesimal amount of work done by $\mathbf{F}$ through the displacement $d\mathbf{r}$. We point out that in this example, the force covector and the displacement vector have different units and even belong to different vector spaces. Nevertheless, their contraction is well defined because one is a vector and the other is a covector.

Example 4 Three-dimensional Euclidean space.

The Euclidean metric on $\mathbb{R}^3$ is

$$ds^2 = dx^2 + dy^2 + dz^2 = \delta_{ij} dx^i dx^j,$$

where $j = 1, 2, 3$ and

$$\delta_{ij} = \delta^{ij} = \delta^i_j = \begin{Bmatrix} 1\ , \ i = j \\ 0\ , \ i \neq j \end{Bmatrix} \tag{3.54}$$

is the *Kronecker delta symbol.* For $x = (x^1, x^2, x^3)$, $y = (y^1, y^2, y^3) \in \mathbb{R}^3$, the *Euclidean scalar product* $x \circ y$ is defined by

$$x \circ y = x^1 y^1 + x^2 y^2 + x^3 y^3 = \delta_{ij} x^i y^j. \tag{3.55}$$

Let $f \in V$ and define a covector $f^* : V \to \mathbb{R}$ by

$$f^*(x) = f \circ x = \delta_{ij} f^i x^j. \tag{3.56}$$

We compute the components of f in the standard basis. Thus,

$$f_i = f^*(\mathbf{e}_i) = f \circ \mathbf{e}_i = f^i.$$

Thus, the vector f and the corresponding covector f^* have the same components. In Euclidean space, the distinction between vector and covector is blurred. Thus, one may freely raise and lower indices on a 3D vector, as long as the ensuing results do not contradict any physics. In Minkowski space, of course, we must distinguish between vectors and covectors.

3.6 Relativistic Energy-Momentum

Energy-momentum plays an important role in physics because

- the energy-momentum of a collection of objects is the sum of the individual energy-momentum of each object,
- The total energy-momentum before an interaction of objects remains the same after the interaction.

Additivity and conservation of energy-momentum play a major role in describing such interactions.

We show first that in Minkowski space, the geodesics are straight lines $x(t) = c(t, (\mathbf{v}/c)t)$, where the velocity $\mathbf{v}$ is constant. This is equivalent to the conservation of all energy-momenta p_λ^{fr}, where the superscript fr, meaning "free," indicates that the object in question is a free object, free of the influence of fields and media.

Let $x(\sigma)$ be a worldline of an object moving in Minkowski space and let

$$L(x,u) = \sqrt{\eta_{\mu\nu}u^\mu u^\nu}. \tag{3.57}$$

The idea behind $L(x, u)$ is that the infinitesimal distance in Minkowski space from x to $x + \epsilon u$ is approximately $L(x, u)\epsilon$, when ϵ is small.

From (2.31), the unit-free energy-momentum covector is

$$p_\mu(\sigma) = \frac{\eta_{\mu\nu}\frac{dx^\nu}{d\sigma}}{\sqrt{\eta_{\mu\nu}\frac{dx^\mu}{d\sigma}\frac{dx^\nu}{d\sigma}}}. \tag{3.58}$$

In order to simplify the formula for p_μ, we choose σ to be the proper time τ with dimensions of length defined as

$$d\tau^2 = \eta_{\mu\nu}dx^\mu dx^\nu. \tag{3.59}$$

Denote differentiation by τ with a dot: $\dot{x} = \frac{dx}{d\tau}$. Note that with this parametrization, the denominator of (3.58) equals 1, and the unit-free energy-momentum covector becomes

$$p_\mu = p_\mu^{\text{fr}} = \eta_{\mu\nu}\dot{x}^\nu = \dot{x}_\mu, \tag{3.60}$$

which is the μ component of the four-velocity, with a lower index. Thus, in the absence of fields, the derivative $\dot{p}_\mu$ is the four-acceleration a_μ. Now, since the function (3.57) does not depend on x^μ, the Corollary to the Theorem of section 2.3 implies that the four-velocity is constant. Therefore, a stationary worldline in Minkowski space has constant velocity.

The *standard four-momentum* $\tilde{p}$, with dimensions of mass times velocity, is mc times the unit-free energy-momentum:

$$\tilde{p} = mcp^{\text{fr}} = (\tilde{p}_0, -\tilde{\mathbf{p}}) = (mc\gamma, -m\gamma\mathbf{v}) . \tag{3.61}$$

The 3D vector

$$\tilde{\mathbf{p}} = m\gamma(v)\mathbf{v} \tag{3.62}$$

is the called the *relativistic momentum* and is Newtonian momentum $m\mathbf{v}$ times the γ factor, expressing the change of parameter from t to τ. This raises the question: what is the physical meaning of $\tilde{p}_0$? The power series expansion of $c\tilde{p}_0 = mc^2\gamma$ is

$$c\tilde{p}_0 = \frac{mc^2}{\sqrt{1 - \frac{v^2}{c^2}}} = mc^2\left(1 + \frac{1}{2}\frac{v^2}{c^2} + \cdots\right) = mc^2 + \frac{1}{2}mv^2 + \cdots . \tag{3.63}$$

We recognize the second term $1/2mv^2$ as the classical kinetic energy. When the velocity is 0, only the first term remains, which we recognize as the particle's rest energy $E_{\text{rest}} = mc^2$, Einstein's famous formula connecting energy and mass. Thus, $\tilde{p}_0$ can be identified with E/c, and

$$E = mc^2\gamma(v) \tag{3.64}$$

is the *total energy* of a particle of mass m moving in Minkowski space. This justifies calling the p_λ the *energy-momenta*. The difference $E - mc^2$ of the total energy and the rest energy is called the *relativistic kinetic energy* K. Thus,

$$K = mc^2(\gamma(v) - 1). \tag{3.65}$$

The definitions of relativistic momentum and relativistic kinetic energy are standard (see, for example, [87, Chap. 6]).

As another example of the conservation of energy-momentum, consider the problem of colliding spheres of masses m and M ($m < M$). We consider two types of collisions: 1) an *elastic collision*, in which the total kinetic energy is conserved, and 2) a *completely inelastic collision*, in which kinetic energy is lost in bonding the two objects together.

Refer now to Fig. 3.12. In the top left of the figure, we see two spheres before a collision (state 0)), in an inertial frame K in which the heavy object B of mass M is at rest, and the lighter object A of mass m has velocity v_0, which is towards B. On the right side, we see the situation in the inertial frame K', whose velocity in K is v_0. In K', A is at rest, and B has velocity $-v_0$, which is towards A.

The middle row of the figure, state 1), shows the situation after an elastic collision, both in K and K'. In K, the velocities after the collision are denoted v_A and v_B. The corresponding velocities in K' denoted $\tilde{v}_A$ and $\tilde{v}_B$. The bottom row, state 2), shows the situation after a completely inelastic collision. The velocity of the combined object is denoted v_{AB} and $\tilde{v}_{AB}$ in K and K', respectively.

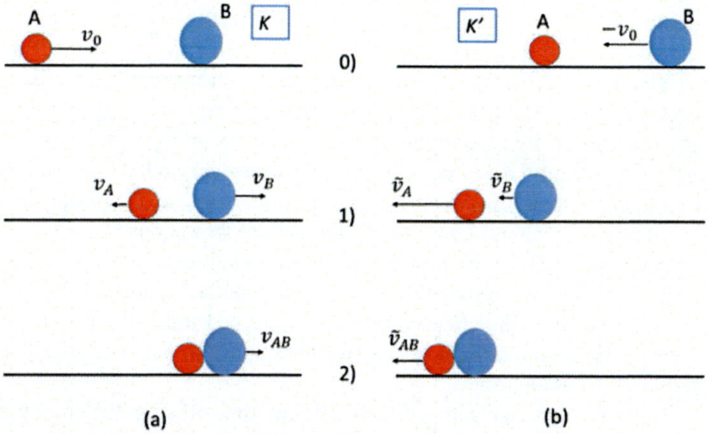

Fig. 3.12 Energy-momentum conservation in a 1D collision of two masses. The three states: (0) before the collision, (1) after an elastic collision, (2) after a completely inelastic collision are displayed **a** in frame K, in which the heavy object B was initially at rest, and **b** in frame K' in which the light object A was initially at rest

The problem is to compute v_A and v_B (or v_{AB}) from the masses m, M and the initial condition v_0. We start with the Newtonian mechanical derivation. A relativistic treatment appears afterwards.

For the elastic collision, conservation of momentum before and after the collision implies that, in K,

$$mv_0 = mv_A + Mv_B.$$

Since kinetic energy is conserved, we also have, in K,

$$mv_0^2/2 = mv_A^2/2 + Mv_B^2/2.$$

The solution of this system of two linear equation is

$$v_A = \frac{m-M}{m+M}v_0, \quad v_B = \frac{2m}{m+M}v_0. \tag{3.66}$$

For the completely inelastic collision, momentum conservation gives

$$mv_0 = (m+M)v_{AB},$$

implying that $v_{AB} = \frac{m}{m+M}v_0$. The reduction in the kinetic energy before and after the collision is

$$\Delta E = \frac{mv_0^2}{2} - \frac{(m+M)v_{AB}^2}{2} = \frac{mM}{m+M}\frac{v_0^2}{2}.$$

Since total energy is conserved, the lost kinetic energy is used to bond the two bodies together.

The above collisions could be observed in the inertial frame K', in which the object A was at rest before the collision. The velocities $\tilde{v}_A, \tilde{v}_B$ are obtained from the above formulas by exchanging m and M and replacing v_0 by $-v_0$. We observe that $v_A = \tilde{v}_A - v_0$ and $v_B = \tilde{v}_B - v_0$, which corresponds to the Galilean transformations of velocities (3.14) between K' and K. This is an expression of the invariance of Newtonian mechanics under the Galilean transformations.

Next, we analyze the problem using the relativistic energy-momentum. In relativity, we use the conservation of the total standard four-momentum $\tilde{p}$, defined by (3.61), instead of the conservation of energy and momentum separately. Since we have restricted the motion to one spatial dimension, our energy-momenta will have only two components.

Consider first an elastic collision. Before the collision, the energy-momentum of A is $mu_0 = m(u_0^0, u_0^1) = m(\gamma_0, \gamma_0\beta_0)$ and that of B is $M(1,0)$. After the collision, the energy-momentum of A is $m(u_A^0, u_A^1) = m(\gamma_A, \gamma_A\beta_A)$ and of B is $M(u_B^0, u_B^1) = M(\gamma_B, \gamma_B\beta_B)$. Thus, the energy-momentum conservation is

$$m(u_0^0, u_0^1) + (M, 0) = m(u_A^0, u_A^1) + M(u_B^0, u_B^1). \tag{3.67}$$

To solve this equation, we split into components and use the fact that for a four-velocity $u = (u^0, u^1)$, we have $(u^0)^2 = 1 + (u_0^1)^2$. Hence,

$$\begin{aligned} mu_0^1 &= mu_A^1 + Mu_B^1 \\ m\sqrt{1+(u_0^1)^2} + M &= m\sqrt{1+(u_A^1)^2} + M\sqrt{1+(u_B^1)^2}. \end{aligned} \tag{3.68}$$

From the first equation, we have $u_B^1 = \frac{m}{M}\left(u_0^1 - x\right)$, where $x = u_A^1$. Substituting this into the second equation, we obtain the following equation, which may be solved numerically:

$$m\sqrt{1+(u_0^1)^2} + M = m\sqrt{1+x^2} + M\sqrt{1+\left(\frac{m}{M}\left(u_0^1 - x\right)\right)^2}. \tag{3.69}$$

After obtaining a value for $x = u_A^1$, the velocity v_A is computed using the relationship

$$v = \frac{cu^1}{\sqrt{1+(u^1)^2}}. \tag{3.70}$$

For a completely inelastic collision, the energy-momentum conservation is

$$m(u_0^0, u_0^1) + (M, 0) = (m+M)(u_{AB}^0, u_{AB}^1) + (\Delta E/c^2, 0). \tag{3.71}$$

Table 3.1 Newtonian and relativistic 1D collisions of two objects of masses m and $M = 2m$. Initially, the velocity of m is $c\beta_0$, and M is at rest. If the collision is elastic, m has velocity $c\beta_A$ and M has velocity $c\beta_B$. If the collision is completely inelastic, the combined object has velocity $c\beta_{AB}$, and the loss of energy in the collision is ΔE

Dynamics	β_0	β_A	β_B	β_{AB}	ΔE (mc^2)
Newtonian	0.3	−0.1	0.2	0.1	0.03
Relativistic	0.3	−0.102	0.204	0.104	0.047
Newtonian	0.6	−0.2	0.4	0.2	0.12
Relativistic	0.6	−0.22	0.438	0.242	0.158
Newtonian	0.9	−0.3	0.6	0.3	0.27
Relativistic	0.9	−0.400	0.781	0.567	0.652

Equating second components, we obtain $u^1_{AB} = \frac{m}{m+M}u^1_0$, and v_{AB} is obtained from (3.70). The first component gives the energy loss

$$\Delta E/c^2 = mu^0_0 + M - \sqrt{(m+M)^2 + m^2(u^1_0)^2}. \tag{3.72}$$

This energy is transferred from kinetic energy to the bonding energy.

To compare the above results for elastic and inelastic collisions, and for Newtonian and relativistic dynamics, see Table 3.1. We observe that relativistic effects become significant only when the initial velocity is close to the speed of light.

For massless particles ($m = 0$), formula (3.61) gives a nonzero energy-momentum only if $\gamma = \infty$, which is equivalent to $v = c$. Indeed, it is known that such particles do have nonzero energy-momentum and move with the speed of light c. This implies that the energy-momentum of a massless particle should be a null vector.

To define the energy-momentum of a photon, recall that every photon has a well-defined *frequency* ν, as well as a 3D *wave vector* $\mathbf{k}$ in the direction of the propagation of the photon. Therefore, we can represent a photon as a plane wave (more precisely, a plane wave packet) with frequency ω moving in the direction $\mathbf{n}$, where $\mathbf{k} = k\mathbf{n}$. The scalar k is called the *wave number*. Without loss of generality, such a wave can be represented as

$$\psi(x) = \cos(\omega t - k\mathbf{n}\cdot\mathbf{x}). \tag{3.73}$$

The expression $\omega t - k\mathbf{n}\cdot\mathbf{x}$ is called the *phase*. From the periodicity of cosine, it follows that $\omega T = 2\pi$, where $T = 1/\nu$ is the period of the wave. Thus, $\omega = 2\pi\nu$ and is called the *angular frequency*. Now consider x as a function of t. The *phase velocity* of the wave is determined by setting the phase equal to 0. Thus, $k\mathbf{n}\cdot\mathbf{x} = \omega t$, or $k\mathbf{n}\cdot\mathbf{v} = \omega$. Since $\mathbf{n}\cdot\mathbf{v} = c$, we have $k = \omega/c$. This implies that the four-covector $(\omega/c, \mathbf{k})$ is null. Using Planck's formula $E = h\nu$, where h is Planck's constant, we define the *energy-momentum* $\tilde{p}$ of a photon as

$$\tilde{p} = \hbar(\omega/c, \mathbf{k}), \tag{3.74}$$

where $\hbar$ is the reduced Planck's constant ($\hbar = 1.054571817... \times 10^{-34}$J · s). Thus, the energy and the momentum are

$$E = \hbar\omega = h\nu, \quad \mathbf{p} = \hbar\mathbf{k}. \tag{3.75}$$

3.7 Relativistic Doppler Shift

In this section, we use of the conservation of energy-momentum to derive the *relativistic Doppler shift*. We follow the derivation of [14]. Consider radiation in the form of a photon emitted from an atom moving with respect to an observer. We assume that the observer is at rest in an inertial system K, that the observer is located on the positive x axis, and that the photon is emitted at the origin O of K. We also assume that the motion of the atom and the photon are in the x, y plane (see Fig. 3.13).

Denote by $\mathbf{v}_1$ and $\mathbf{v}_2$, respectively, the velocities of the atom before and after the emission of the photon. Denote their angles with the x-axis by ϕ_1 and ϕ_2, respectively. We denote the rest energy of the atom before emission by E_1 and its rest energy after the emission by E_2. If the atom was at rest during the emission, then by conservation of energy, the energy of the emitted photon is the energy difference $E_1 - E_2$, implying by (3.75) that its frequency ν_0 satisfies

$$h\nu_0 = E_1 - E_2. \tag{3.76}$$

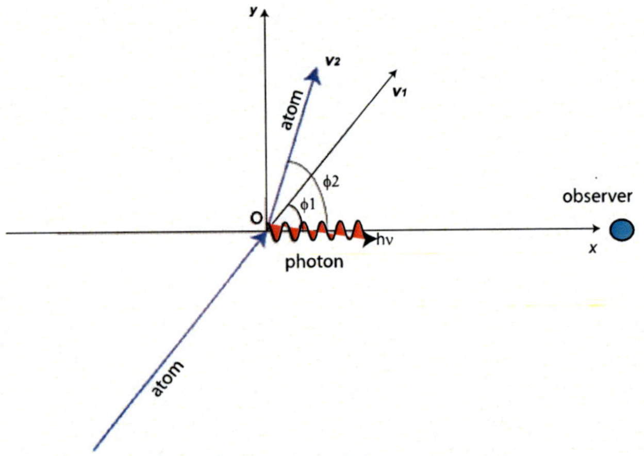

Fig. 3.13 Photon emission and the Doppler shift

From (3.61), it follows that for an atom moving with velocity $\mathbf{v}_i, i = 1, 2$, the energy-momentum of the atom in our lab frame is $(mc\gamma_i, m\gamma_i\mathbf{v}_i)$, where $\gamma_i = \gamma(v_i) = 1/\sqrt{1 - v_i^2/c^2}$. By Planck's formula, the energy of the emitted photon is $h\nu$, where ν is the observed frequency, and its 3D momentum is in the x direction and is $p_x = \hbar k = h\nu/c$.

Energy conservation before and after the emission leads to

$$\gamma_1 E_1 = \gamma_2 E_2 + h\nu, \tag{3.77}$$

and, using $E_i = m_i c^2$, momentum conservation in the x and y directions yields

$$\gamma_1 \frac{E_1}{c^2} v_1 \cos\phi_1 = \gamma_2 \frac{E_2}{c^2} v_2 \cos\phi_2 + \frac{h\nu}{c} \tag{3.78}$$

and

$$\gamma_1 \frac{E_1}{c^2} v_1 \sin\phi_1 = \gamma_2 \frac{E_2}{c^2} v_2 \sin\phi_2. \tag{3.79}$$

Subtracting c(3.78) from (3.77) and introducing $\beta_i = v_i/c$, we obtain

$$\gamma_1 E_1 (1 - \beta_1 \cos\phi_1) = \gamma_2 E_2 (1 - \beta_2 \cos\phi_2). \tag{3.80}$$

Introduce

$$\varphi_i = \gamma_i (1 - \beta_i \cos\phi_i), \quad \psi_i = \gamma_i \beta_i \sin\phi_i.$$

Then, using that $\gamma_i^2 = 1/(1 - \beta_i^2)$, one obtains $1 + \varphi_i^2 + \psi_i^2 = 2\gamma_i\varphi_i$, and

$$\gamma_i = \frac{1 + \varphi_i^2 + \psi_i^2}{2\varphi_i}.$$

From (3.80) and (3.79), we obtain

$$E_1\varphi_1 = E_2\varphi_2 = a, \quad E_1\psi_1 = E_2\psi_2 = b$$

and

$$E_i\gamma_i = \frac{E_i^2 + (E_i\varphi_i)^2 + (E_i\psi_i)^2}{2E_i\varphi_i} = \frac{E_i^2 + a^2 + b^2}{2a}.$$

Using (3.77), we now obtain

$$h\nu = E_1\gamma_1 - E_2\gamma_2 = \frac{1}{2a}(E_1^2 - E_2^2).$$

Finally, using (3.76), we obtain the *true relativistic Doppler shift formula*

$$\nu = \frac{E_1 + E_2}{2E_1\gamma_1(1 - \beta_1 \cos\phi_1)}\nu_0.$$

The commonly used relativistic Doppler formula is

$$\nu_0 = \gamma(v)(1 - \beta\cos\phi)\nu,$$

but is valid only as an approximation when $h\nu_0 \ll E_2 < E_1$ which implies that $E_1 \approx E_2 = E$, $v_1 \approx v_2 = v$, and $\phi_1 \approx \phi_2 = \phi$.

3.8 Lorentz-Covariant Functions for a Single-Source Field

Our action function (see Introduction, formula (1.1)) is built from four-covector-valued functions of spacetime position. In order to satisfy the Principle of Relativity, these functions should be Lorentz covariant, as defined below. We show here, for a single-source field, how to construct Lorentz-covariant four-vector-valued functions. By lowering indices, one obtains Lorentz-covariant four-covector-valued functions.

We consider the Lorentz covariance of functions $f_{\psi(\tau)}(x)$ which depend on the worldline $\psi(\tau)$ of a given source. The notion of a *Lorentz-covariant four-vector-valued function* is then defined as follows. Let K and K' be two arbitrary inertial frames. Let M and M', respectively, be the collection of coordinates of spacetime positions in K and K', respectively. Let $\Lambda : M \to M'$ be the Poincaré transformation from M to M'. Let $\psi(\tau)$ be a worldline in K, and let $\psi'(\tau)$ be the worldline $\Lambda\psi(\tau)$ in K'. We say that the function $f_{\psi(\tau)}(x) : M \to \tilde{M}$ is *Lorentz covariant* if, for any spacetime point x in K, the Poincaré transformation Λ maps $f_{\psi(\tau)}(x)$ to the same four-vector as $f_{\psi'(\tau)}$ maps Λx. Mathematically, this can be expressed by saying that the following diagram commutes:

$$\begin{array}{ccc} M & \xrightarrow{\;f_{\psi(\tau)}\;} & \tilde{M} \\ {\scriptstyle\Lambda}\downarrow & & \downarrow{\scriptstyle\Lambda} \\ M' & \xrightarrow{\;f_{\psi'(\tau)}\;} & \tilde{M}' \end{array}$$

That is, $f_{\psi(\tau)}$ is Lorentz covariant if

$$\Lambda f_{\psi(\tau)}(x) = f_{\psi'(\tau)}(\Lambda x) \tag{3.81}$$

for any spacetime point $x \in M$ and any Poincaré transformation on M.

Consider now a field (gravitational or electromagnetic) of a single source with worldline $\psi(\tau)$ (Fig. 3.14). Let P be an arbitrary spacetime position, with coordinates x. Define the retarded time $\tau(x)$ such that $(x - \psi(\tau(x)))^2 = 0$ and $x^0 - \psi^0(\tau(x)) > 0$. The point Q with coordinates $\psi(\tau(x))$ is the unique intersection of the backward

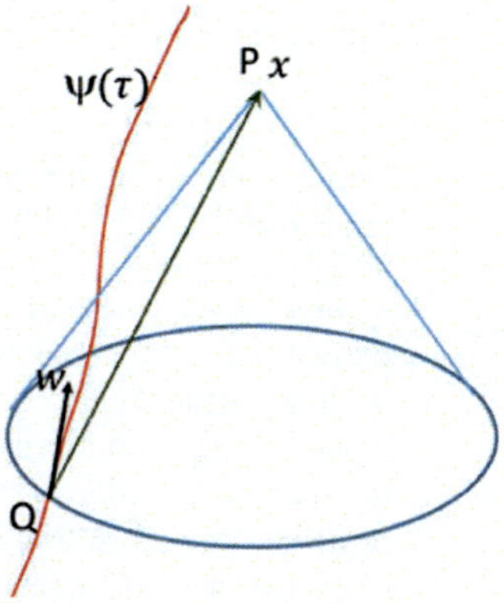

Fig. 3.14 The point Q is the unique intersection of the worldline $\psi(\tau)$ of the source and the backward light cone at P. The four-velocity of the source at the retarded time is $w(\tau(x))$ and is tangent to the worldline of the source at the point Q

light cone with vertex x and the worldline $\psi(\tau)$ of the source. Since the field propagates with the speed of light, the position of the source at the retarded time is the only influence on the object at x. Thus, the relative position four-vector PQ is a preferred direction from the point P, where we wish to describe the field, to the preferred point Q representing the source at the retarded time. Hence, at each point x, there is a four-vector

$$r(x) = x - \psi(\tau(x)), \tag{3.82}$$

which we call the *relative position vector* of x with respect to the source. From this vector, we construct Lorentz-covariant four-vector-valued functions.

We check now that our definition of the four-vector $r(x)$ is Lorentz covariant. Let Λ be a Poincaré transformation, let $x' = \Lambda x$, and let $\psi' = \Lambda\psi$. Since Λ preserves the norm, it maps the null four-vector $r(x)$ to a null four-vector r'. Moreover, since r is a four-vector, Eq. (3.5) implies that Λ acts linearly on r. Hence,

$$r' = \Lambda r = \Lambda(x - \psi(\tau(x))) = \Lambda x - \Lambda\psi(\tau(x)) = x' - \Lambda\psi(\tau(x)).$$

Now, since $r' = x' - \Lambda\psi(\tau(x))$ is null, the point $\Lambda\psi(\tau(x))$ is on the backward light cone at x'. This means that Λ maps the intersection point Q to the intersection point Q' of the backward light cone at x' and the worldline ψ'. Therefore, the four-vector r' at the point Λx is precisely the image under Λ of the four-vector r at the point x. Eq. (3.81) thus holds, meaning that r is defined in a Lorentz-covariant way.

In addition to the four-vector $r(x)$, one may also use the four-velocity $w(\tau(x))$ of the source at the retarded time to construct Lorentz-covariant four-vector-valued functions. Moreover, the inner product $r \cdot w$ is a Lorentz-invariant scalar, as is any scalar function of $r \cdot w$. Thus, at each point x, there are two types of Lorentz-covariant four-vector-valued functions:

$$f(x) = g(r(x) \cdot w(\tau(x)))r(x) \tag{3.83}$$

and

$$f(x) = h(r(x) \cdot w(\tau(x)))w(\tau(x)), \tag{3.84}$$

where g and h are scalar functions of the Lorentz-invariant scalar $r \cdot w$. We will use the Newtonian limit to identify these functions explicitly, in Chap. 5 for an electromagnetic field, and in Chap. 6 for a gravitational field.

In the case of motion in isotropic media, there are no preferred points and no preferred directions. Hence, any Lorentz-covariant four-covector-valued function must be constant.

In passing, we note that the *four-acceleration* $a = dw/d\tau$ is also a four-vector and thus may be used to construct Lorentz-covariant functions. For the sake of simplicity, however, we will restrict ourselves to functions of the forms (3.83) and (3.84).

In the context of the paragraph before Eq. (3.83), we compute now the first-order derivatives of the retarded time $\tau(x)$, the relative position null four-vector $r(x) = x - \psi(\tau(x))$, the source four-velocity $w(\tau)$, and the inner product $r \cdot w$. For ease of notation, we differentiate by the parameter τ. We will need these derivatives, beginning in Chap. 5.

The partial derivative $r_{,\mu} = \partial r/\partial x^\mu$ is

$$r^\alpha_{,\mu} = x^\alpha_{,\mu} - \psi^\alpha_{,\mu}(\tau(x)) = x^\alpha_{,\mu} - \partial\psi^\alpha/\partial\tau\, \tau_{,\mu} = x^\alpha_{,\mu} - w^\alpha \tau_{,\mu}. \tag{3.85}$$

Note that $x^\alpha_{,\mu} = \delta^\alpha_\mu$. The vector $r(x)$ is null, implying that $r \cdot r_{,\mu} = 0$. Thus,

$$0 = r \cdot r_{,\mu} = r_\alpha r^\alpha_{,\mu} = r_\alpha \delta^\alpha_\mu - r_\alpha w^\alpha \tau_{,\mu} = r_\mu - (r \cdot w)\tau_{,\mu}.$$

Hence, the derivative of the retarded time is

$$\tau(x)_{,\mu} = \frac{r_\mu}{r \cdot w}, \tag{3.86}$$

and (3.85) becomes

$$r^\nu_{,\mu} = \delta^\nu_\mu - \frac{w^\nu r_\mu}{r \cdot w}. \tag{3.87}$$

We compute $r_{\nu,\mu}$ by lowering the index:

$$r_{\nu,\mu} = (\eta_{\nu\alpha} r^\alpha)_{,\mu} = \eta_{\nu\alpha} r^\alpha_{,\mu} = \eta_{\nu\alpha}\left(\delta^\alpha_\mu - \frac{w^\alpha r_\mu}{r \cdot w}\right) = \eta_{\mu\nu} - \frac{w_\nu r_\mu}{r \cdot w}. \tag{3.88}$$

For the derivatives of the four-velocity of the source, we have

$$w_{\nu,\mu} = \partial_\mu w(\tau(x))_\nu = \partial w_\nu/\partial\tau\, \tau_{,\mu} = a_\nu(\tau)\frac{r_\mu}{r \cdot w} = \frac{a_\nu r_\mu}{r \cdot w}, \tag{3.89}$$

where $a_\nu(\tau(x))$ is the source four-acceleration covector at the retarded time. Similarly, using that $(r)^2 = 0$ and $(u)^2 = 1$, we obtain

$$(r \cdot w)_{,\mu} = (r^\nu w_\nu)_{,\mu} = r^\nu_{\ ,\mu} w_\nu + r^\nu w_{\nu,\mu} = w_\mu - \frac{r_\mu}{r \cdot w} + \frac{r_\mu (a \cdot r)}{r \cdot w}. \tag{3.90}$$

Note that $r \cdot w$ is the spatial distance from the point P where we wish to compute the field to the source at the retarded time in the inertial frame K' comoving to the source at the retarded time. This is because r is null and, in K', $w = (1, 0, 0, 0)$. We also point out that if the null vector r is replaced by $2r$, then the last term of $(r \cdot w)_{,\mu}$ will double if the source is accelerating. On the other hand, the derivatives $r_{\nu,\mu}$ and $w_{\nu,\mu}$ will not change.

Chapter 4
The Geometric Model of Relativistic Dynamics

In this chapter, we present the new ideas which form the foundation of our novel approach to relativistic dynamics. Our approach is geometric and describes the motion of massive and massless, charged and uncharged objects under the influence of force fields and isotropic media. For the time being (until Chap. 8), we consider only the trajectory of the object, considered as a point, and ignore any internal rotation.

We introduce the *relativity of spacetime*. This is a new idea and one of the pillars of our dynamics. It is the notion that each object has its own spacetime. Moreover, according to the *Extended Principle of Inertia*, another new idea [42], an object moves along a path which is stationary with respect to the geometry of *its* spacetime. An object's spacetime is determined by the forces affecting it. To describe the geometry of this spacetime, we introduce a simple and physically meaningful action function. Using this action function, which first appeared in [48], we obtain a universal relativistic dynamics equation which incorporates gravity, electromagnetism and isotropic media.

4.1 The Relativity of Spacetime and the Extended Principle of Inertia

On August 2, 1971, Commander David Scott simultaneously dropped a hammer and a feather. With the entire world watching, the hammer and the feather hit the surface of the moon at the same time. This simple yet dramatic demonstration validated Galileo's four-hundred-year-old claim that bodies falling in a vacuum under the influence of gravity fall at the same rate, regardless of their mass. Thus, gravity is an object-independent force. As a result, it lends itself to being modelled by geometry. Indeed, the first application of Riemann's idea that "force equals geometry" was

Y. Friedman and T. Scarr, *A Novel Approach to Relativistic Dynamics*,
Fundamental Theories of Physics 210,
https://doi.org/10.1007/978-3-031-25214-3_4

Chapter 4
The Geometric Model of Relativistic Dynamics

In this chapter, we present the new ideas which form the foundation of our novel approach to relativistic dynamics. Our approach is geometric and describes the motion of massive and massless, charged and uncharged objects under the influence of force fields and isotropic media. For the time being (until Chap. 8), we consider only the trajectory of the object, considered as a point, and ignore any internal rotation.

We introduce the *relativity of spacetime*. This is a new idea and one of the pillars of our dynamics. It is the notion that each object has its own spacetime. Moreover, according to the *Extended Principle of Inertia*, another new idea [42], an object moves along a path which is stationary with respect to the geometry of *its* spacetime. An object's spacetime is determined by the forces affecting it. To describe the geometry of this spacetime, we introduce a simple and physically meaningful action function. Using this action function, which first appeared in [48], we obtain a universal relativistic dynamics equation which incorporates gravity, electromagnetism and isotropic media.

4.1 The Relativity of Spacetime and the Extended Principle of Inertia

On August 2, 1971, Commander David Scott simultaneously dropped a hammer and a feather. With the entire world watching, the hammer and the feather hit the surface of the moon at the same time. This simple yet dramatic demonstration validated Galileo's four-hundred-year-old claim that bodies falling in a vacuum under the influence of gravity fall at the same rate, regardless of their mass. Thus, gravity is an object-independent force. As a result, it lends itself to being modelled by geometry. Indeed, the first application of Riemann's idea that "force equals geometry" was

Y. Friedman and T. Scarr, *A Novel Approach to Relativistic Dynamics*,
Fundamental Theories of Physics 210,
https://doi.org/10.1007/978-3-031-25214-3_4

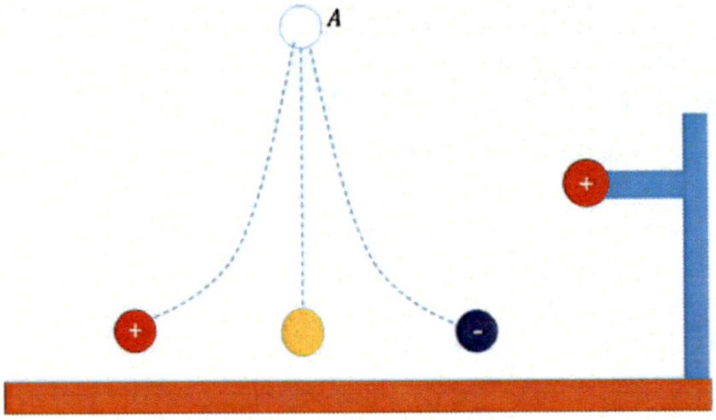

Fig. 4.1 On the right is a positive charge, fixed in position. A negative charge, moving freely at the point A, is accelerated toward the fixed positive charge. A positive charge is accelerated away from the fixed positive charge. A neutral particle is not influenced by the positive charge and continues to move freely

GR. In Einstein's theory, mass curves spacetime and, by the Equivalence Principle, mass curves spacetime in the same way for everybody. Curved spacetime can thus be considered a *stage* on which all objects move. In other words, the geometry of spacetime is the same for all objects. That's why the hammer and the feather landed simultaneously.

Now, what happens when we try to use the above physics to describe the electromagnetic force? Take the simplest possible case: place a positive charge Q at some fixed point, as in Fig. 4.1. At some other point nearby, place three test charges, one positive, one negative, and one neutral. What happens? The positive test charge accelerates away from Q, the negative test charge accelerates toward Q, and the neutral particle moves uniformly, unaffected by Q. Clearly, the electromagnetic field does *not* create a common stage on which all particles (charged and uncharged) move. There's not even a single geometry that applies to all *charged* particles. It is well known, in fact, that the acceleration of a charged particle in an electromagnetic field depends on the particle's charge-to-mass ratio, a property which is intrinsic to the object in motion. Electromagnetism, unlike gravity, is an object-dependent force.

Let us consider an additional example. In Fig. 4.2, white light enters a prism and is dispersed into its constituent colors, producing a rainbow. The angle by which a light ray bends upon entering and exiting a prism is determined by Snell's Law and depends on the velocity of the light in the prism (this velocity is typically less than c, the speed of light in vacuum) and the incident angle of the light with the prism. But the amount of bending also depends on the *frequency* of the light, a property intrinsic to the light. Different colors (frequencies) are bent by different angles. Like the motion of charges in an electromagnetic field, the motion of light in media is object dependent.

These examples raise the question: Can Riemann's principle of "force equals geometry" be applied to object-dependent forces? If a force doesn't create a common stage, then how can a geometric description even get off the ground? We show in this chapter how to circumvent these obstacles and describe object-dependent forces with geometry. To accomplish this, we introduce two new ideas which form the foundation of our approach to relativistic dynamics: the *relativity of spacetime* and the *Extended Principle of Inertia*.

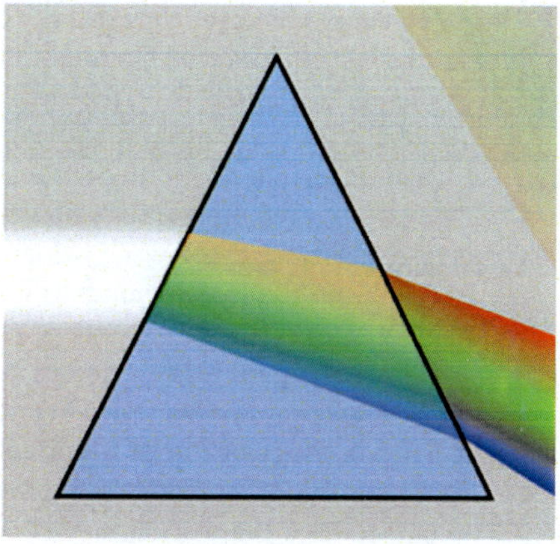

Fig. 4.2 The relativity of light propagation. A prism disperses white light into its constituent colors. This demonstrates that photons of different frequency have different spacetimes. Used with permission from Creative Commons under the license https://creativecommons.org/licenses/by-sa/3.0/deed.en. The original image was cropped

The *relativity of spacetime* is the notion that spacetime is an object-dependent notion. *Each object has its own spacetime*. A positive charge has its own spacetime, and its spacetime is different from that of a negative charge. The spacetime of red light is different from that of blue light.

What determines an object's spacetime? An object's spacetime is determined by the forces which affect its motion and at most one parameter intrinsic to the object. For example, a massive, non-charged object's spacetime is defined by gravity and nothing else, not even its mass. The source masses curve spacetime in a way that is independent of the mass of the object moving in the field. In GR, gravity is "defined away" from a force and into a common stage on which all objects move.

Now if gravity were the only force of nature, then every object's spacetime would be the same. Spacetime wouldn't be relative. But there *are* other forces. Spacetime *is* relative. A charged particle, for the example, is affected by both gravity and electromagnetism. Its spacetime is defined by the source masses, the source charges, and its own charge-to-mass ratio. Inside isotropic media, the motion of photons and charges are determined by the properties of the media and at most one parameter intrinsic to the particle. In the prism example above, this parameter is the photon's wavelength. In this book, we consider spacetimes defined by gravitational fields, electromagnetic fields, and isotropic media. Table 4.1 summarizes the various ingredients in different particles' spacetimes.

Table 4.1 The factors determining an object's spacetime. A ✓ indicates that the given force or influence affects the given particle. n/a = not addressed in this book

	Gravity	Electromagnetism	Media	Intrinsic Property
massive, uncharged	✓		n/a	
massive, charged	✓	✓	✓	✓
photon	✓		✓	✓

Having established the relativity of spacetime, the question is: how does an object moves in its spacetime? This question is answered by the next new idea: the *Extended Principle of Inertia*. To understand how, we must return, for the moment, to the (original) Principle of Inertia.

The Principle of Inertia implies that the worldline of free motion is a straight line in the inertial observer's spacetime. This line is stationary with respect to the Minkowski metric. Now *why* does an object *not* move along a straight worldline when there are forces around? Is it because there are forces around, or because our definition of "straight" is wrong? Before Riemann, the answer was "because there are forces around." These forces cause the object's worldline to *curve*. Riemann, however, suggested that our definition of "straight" is wrong. The forces cause a change in the geometry of space, and in this altered geometry, the object's worldline is "straight." It is a *geodesic*, a curve of shortest distance between any two points on it. A geodesic is a worldline of constant velocity or zero acceleration, as defined in the geometry of the object's spacetime.

This idea was implemented by Einstein. In GR, an object under the influence of a gravitational field moves along a geodesic in a curved spacetime. The curving of the spacetime is defined by the sources of the field. In GR, one extends the definition of "straight" to include these geodesics. What appears curved to an inertial observer is a straight line *in the object's spacetime*. In GR, objects move along "straight" lines *even when there are gravitational forces around*. In this book, we show how to implement these ideas not only for gravity, but also for electromagnetism and motion in isotropic media.

Motion along a geodesic means motion with *constant velocity*. An object moves with constant velocity because it has no internal mechanism, like an accelerator or a steering wheel, with which to change its velocity. This is clear when there are no forces around. But when there *are* forces around, the object still has no way to change its velocity *of its own accord*. Hence, it still moves along a geodesic, along a "straight" or stationary worldline in its own spacetime. However, since we want to include non-gravitational forces as well, we cannot assume a priori that the curving is independent of the moving object. We allow the object's spacetime to depend on (at most) one property intrinsic to the object.

These ideas are encapsulated in the **Extended Principle of Inertia**:

Since an inanimate object (not disturbed by other objects) is unable to change its velocity, its worldline is stationary in its spacetime.

The Extended Principle of Inertial implies that every object travels along a geodesic with respect to the geometry of the object's spacetime. (By "disturbed," we mean by direct contact, that is, a collision.)

To calculate geodesics, we require a method of measuring the length of an object's worldline in its spacetime. For this purpose, we introduce an *action function*. In the next section, we use the ancient problem of finding the shortest route between two points on the Earth's surface to provide some intuition about the definition of the action function and the use of a flat background for calculations, as well as to introduce our notations. Our method relies on the *Principle of Least Action* and uses the *Euler-Lagrange equations* to find geodesics. The general case begins in Sect. 4.3.

The idea that each inanimate object has its own spacetime can also be applied also to human behavior. True, a human being is *not* an inanimate object. He *does* have mechanisms with which to modify his behavior. He has free will. Nevertheless, as experience testifies, changing one's behavior is challenging. Some people take years to break a habit, if they stop at all. It's much "simpler" to coast through life, riding a geodesic. Others have health issues which they seem unable to resolve. In extreme cases, such as clinical depression, a person can lose or believe they have lost the ability to change. In many of these instances, the problem is that there is one particular influence that dominates this person's world, effectively eclipsing other, more positive influences. A possible course of treatment is to introduce the person to new influences. For example, get a chronically ill person to start exercising. This will be an additional influence defining his world. If a person cannot change his behavior for some reason, we can try to change his world. This will change his geodesic.

For another example, consider psychological projection—the process of attributing one's negative character traits to other people. The Talmud [5] states that when one criticizes another person, he criticizes him for a defect which he himself has. This is the proverbial pot calling the kettle black. From the point of view of our model, the explanation of this phenomenon is simple. A person notices the defect in others because this defect exists in his own world. If he didn't have this flaw, he wouldn't notice it in others.

4.2 Geodesics on the Globe

To help the reader familiarize himself with action functions, stationary worldlines, and working in inertial frames on a flat map, we consider the ancient problem of finding the shortest route between two points P and Q on the Earth's surface, which we take to be a sphere of radius $R = 6.371 \times 10^6$ meters. We will use the standard *geographic coordinates* of latitude ($-\pi/2 \leq \theta \leq \pi/2$) and longitude ($-\pi < \lambda \leq \pi$). The arc $\lambda = 0$ running from the North Pole through Greenwich, England, and on to the South Pole is called the *prime meridian*. It is one half of a great circle as are all meridians $\lambda = \lambda_0$. The great circle $\theta = 0$ is called the *equator*. The circle $\theta = \theta_0 \neq 0$ of constant and nonzero latitude is called a *parallel* and is not a great circle, but a circle of radius $R \cos \theta_0$. A point on the Earth's surface will be represented by an upper case letter and its coordinates, as in $P(\lambda, \theta)$. In addition to the globe, we also use a "flat map," with orthogonal λ and θ axes. As shown in Fig. 4.3. The point on the flat map corresponding to the point $P(\lambda, \theta)$ on the globe will be denoted with a lower case letter, as in $p(\lambda, \theta)$. It is useful to know the coordinate transformation from geographic coordinates to Cartesian coordinates:

$$x = R \cos \lambda \cos \theta\,, \; y = R \sin \lambda \cos \theta\,, \; z = R \sin \theta. \tag{4.1}$$

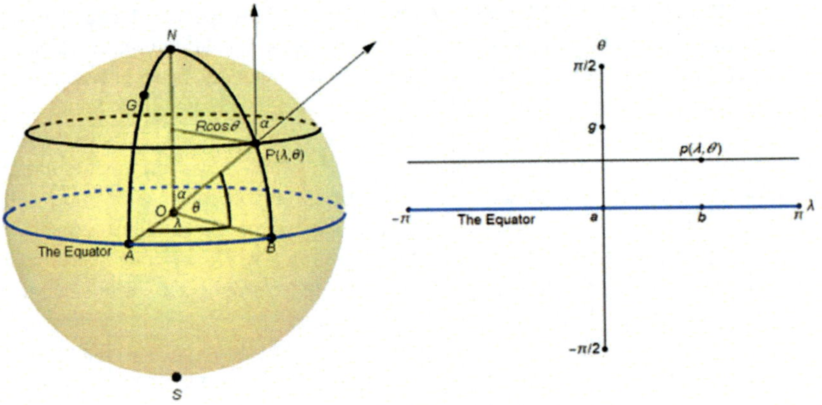

Fig. 4.3 (Left) Geographic coordinates on the Earth's surface. N and S represent, respectively, the North and South Pole. The prime meridian $\lambda = 0$ runs through Greenwich, England, symbolized by G. The longitude λ of a point P is measured from the prime meridian. The latitude θ of P is measured from the equator $\theta = 0$, shown in blue. The center of the Earth is at O. The arc from P to B is part of a circle of radius R and has length $R\theta$. The parallel passing through P is a circle of radius $R\cos\theta$. (Right) Geographic coordinates on a flat map. Longitude is on the horizontal axis, and latitude is on the vertical axis. The points g, p, a and b (lower case letters) correspond, respectively, to the points G, P, A and B on the globe. The equator is shown in blue. The parallel through P is shown in black

Note that it is possible to determine the geographic coordinates of one's location P on the Earth's surface. To define the latitude, measure the angle α (in radians) between the direction to the North Star and the vertical direction (the direction away from the center of the Earth) at P. Since α is also the angle from the center of the Earth between the North Pole and the point P, the latitude $\theta = \frac{\pi}{2} - \alpha$. See Fig. 4.3. To define the longitude, calibrate a clock to read 12 noon when it is midday at Greenwich. Record the time on this clock when it is midday at the point P (which can be identified by the shortest Sun shadow from a vertical pole). Each hour of difference of these two times represents a difference in longitude of $\frac{\pi}{12}$ radians. Thus, it is possible to determine one's geographic coordinates *locally*.

Now, suppose one wants to travel from point P to point Q by the shortest path possible. A path can be defined as a pair of functions $x(\sigma) = (\lambda(\sigma), \theta(\sigma))$, $0 \leq \sigma \leq 1$, with $P = (\lambda(0), \theta(0))$ and $Q = (\lambda(1), \theta(1))$. Subdivide the interval $[0, 1]$ into n subintervals, each of length $\epsilon = 1/n$, and let $x_m = (\lambda(m\epsilon), \theta(m\epsilon))$ (Fig. 4.4). Then the length $S(x)$ of the path is approximately the sum of the lengths from x_m to x_{m+1}, for $m = 0, 1, 2, ..., n - 1$. By choosing n sufficiently large, the problem is reduced to defining the distance between two nearby points $P(\lambda, \theta)$ and $P(\lambda + \Delta\lambda, \theta + \Delta\theta)$.

We will approximate this infinitesimal distance using the flat map. See Fig. 4.5. Introduce a point $p(\lambda + \Delta\lambda, \theta)$. The line from $p(\lambda, \theta)$ to $p(\lambda + \Delta\lambda, \theta)$ represents an arc on the globe which is part of a circle R_θ of radius $R\cos\theta$. The central angle of this arc is $\Delta\lambda$, implying that this arc has length $R\cos\theta\Delta\lambda$. Similarly, the line

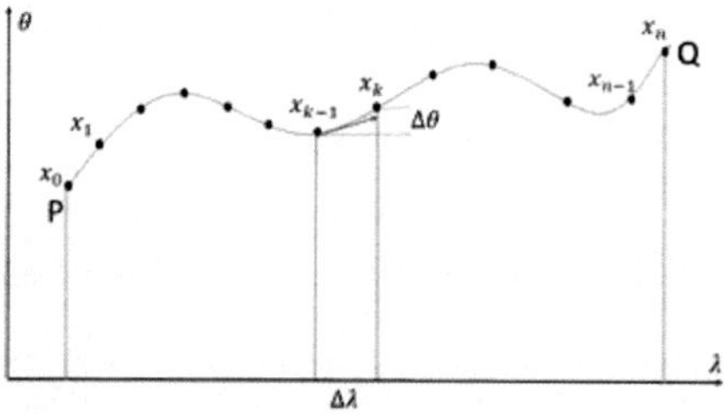

Fig. 4.4 A partition of the path from P to Q into n subintervals. This partition is used to construct a Riemann sum which approximates the length of the worldline. As the number of subintervals approaches infinity, the approximate lengths converge to the actual length

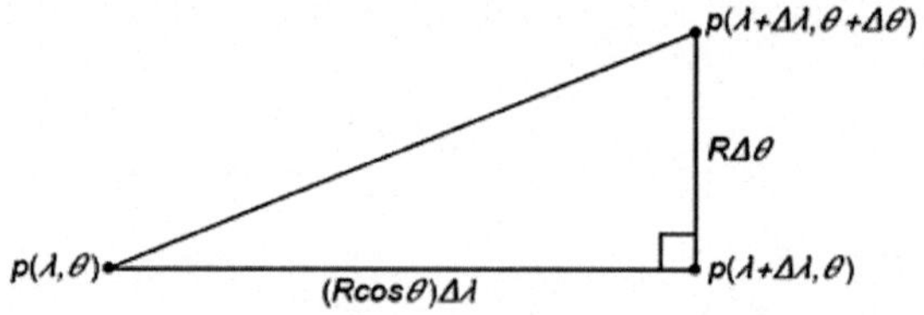

Fig. 4.5 The distance on the globe between two infinitesimally close points, represented on the flat map. The line from $p(\lambda, \theta)$ to $p(\lambda + \Delta\lambda, \theta)$ represents an arc on the globe which is part of a circle of radius $R\cos\theta$. This arc has length $R\cos\theta\Delta\lambda$. The line from $p(\lambda + \Delta\lambda, \theta)$ to $p(\lambda + \Delta\lambda, \theta + \Delta\theta)$ represents an arc on the globe which is part of a (great) circle of radius R. This arc has length $R\Delta\theta$. The distance from $P(\lambda, \theta)$ to $P(\lambda + \Delta\lambda, \theta + \Delta\theta)$ on the globe is approximately $R\sqrt{\Delta\theta^2 + \cos^2\theta\Delta\lambda^2}$

from $p(\lambda + \Delta\lambda, \theta)$ to $p(\lambda + \Delta\lambda, \theta + \Delta\theta)$ represents an arc on the globe which is part of a (great) circle of radius R, with central angle $\Delta\theta$. Its length is thus $R\Delta\theta$. Since the angle between these arcs is approximately 90°, the Pythagorean theorem implies that the distance on the globe between $P(\lambda, \theta)$ and $P(\lambda + \Delta\lambda, \theta + \Delta\theta)$ is approximately

$$D(P(\lambda, \theta), P(\lambda + \Delta\lambda, \theta + \Delta\theta)) \approx R\sqrt{\Delta\theta^2 + \cos^2\theta\Delta\lambda^2}. \tag{4.2}$$

Note that the arcs above are not line segments, but approach line segments when $\Delta\lambda$ and $\Delta\theta$ approach zero.

Fix a point $p(\lambda, \theta)$ on the flat map and a vector $u = (u^1, u^2)$. We define

$$\begin{aligned} L(\lambda, \theta, u^1, u^2) &= \lim_{\epsilon \to 0} \frac{D(P(\lambda, \theta), P(\lambda + \epsilon u^1, \theta + \epsilon u^2))}{\epsilon} \\ &= R\sqrt{(u^2)^2 + \cos^2\theta (u^1)^2}. \end{aligned} \tag{4.3}$$

The definition of L implies that when ϵ is sufficiently small, the distance on the globe between $P(\lambda, \theta)$ and $P(\lambda + \epsilon u^1, \theta + \epsilon u^2)$ is approximately $L(p, u)\epsilon$.

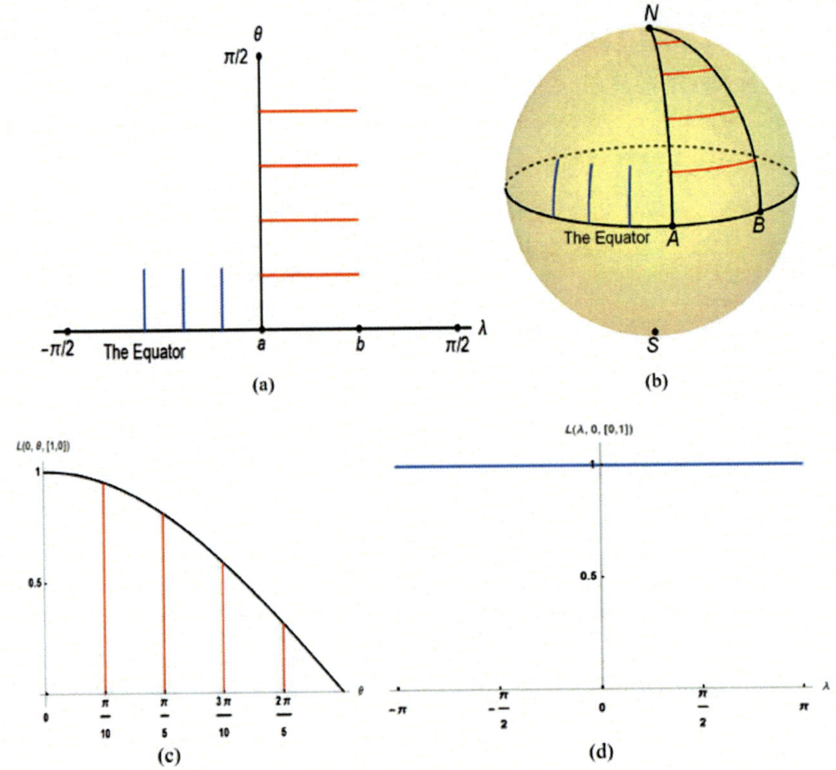

Fig. 4.6 Parallels and meridians on the globe and on the flat map. **a** On the flat map, meridians (arcs of constant longitude λ, shown in blue) that span equal amounts of latitude have equal lengths. Parallels (arcs of constant latitude θ, shown in red) which span equal amounts of longitude also have equal lengths. **b** On the globe, the corresponding meridians have equal lengths, but the parallels become shorter toward the poles. **c** The action function $L(\lambda, \theta, u)$ of Eq. (4.3), for $u = (0, 1)$ parallel to the equator. The length of an interval on the globe is the product of the length on the map and L. Thus, as the latitude θ increases, the intervals become shorter. **d** The action function L for $u = (0, 1)$ parallel to the meridians is constant. Spans of equal latitude have the same length along all meridians

To get a feel for the geometry on the globe, we depict, in Fig. 4.6, lines of constant longitude λ (meridians) and lines of constant latitude θ (parallels). Along a meridian $\lambda = \lambda_0$, we have $u = (0, 1)$, and $L(\lambda_0, \theta, u) = R$ is constant. Along a parallel $\theta = \theta_0$, we have $u = (1, 0)$ and $L(\lambda, \theta_0, u) = R \cos\theta$.

Next, we want to demonstrate the dependence of the action function L on the direction u. For a given point p on the map, the dependence of L on u can be viewed from the image on the globe of a small circle around p. In Fig. 4.7, we display the change of the dependence of L on u for points with different latitudes.

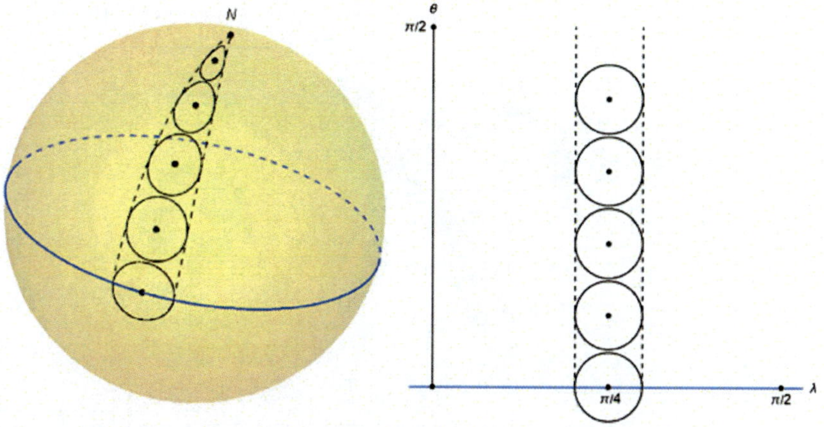

Fig. 4.7 Circles on the flat map and the corresponding curves on the globe. On the flat map (right) are circles of radius 0.15 radians, with respective centers at latitudes 0, $\pi/10$, $\pi/5$, $3\pi/10$ and $2\pi/5$ radians N. On the globe (left), the corresponding curve at the equator is a circle. As the latitude increases, the curves become increasingly egg-shaped. The dashed lines are $\lambda = \pi/4 \pm 0.15$. The equator is in blue

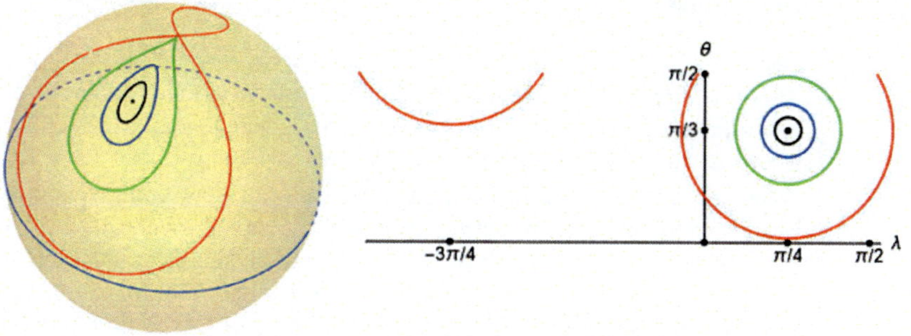

Fig. 4.8 Concentric circles on the flat map and the corresponding curves on the globe. On the flat map are circles with common center $\lambda = \Pi/4$, $\theta = \Pi/3$. The red, green, blue, and black circles have respective radii $1, 0.5, 0.25$ and 0.125 radians. Note that the red circle is composed of two distinct pieces, since it reaches beyond the North Pole. On the globe, the image of the red circle is not convex and crosses itself at the North Pole. As the radius tends to 0, the curves on the globe tend to ellipses. The equator is in blue

Figure 4.8 explains the need of taking the limit $\epsilon \to 0$ in the definition of the action function L. Around the fixed point $\lambda = \Pi/4$, $\theta = \Pi/3$ are concentric circles of different radii. We observe that only for small radii do the images on the globe of these circle approach ellipses, which have a relative simple description. Nevertheless, for the definition of the length of the path, this is all that is needed.

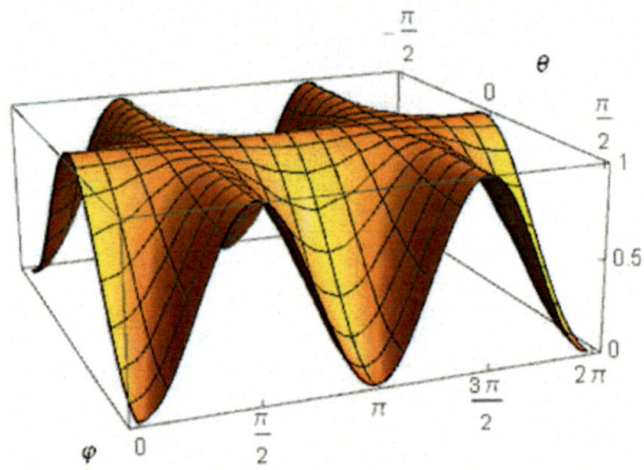

Fig. 4.9 A 3D representation of the action function $L(\lambda, \theta, u) = \sqrt{(u^2)^2 + \cos^2\theta(u^1)^2}$. In the figure, the latitude θ ranges from $-\pi/2$ to $\pi/2$ along the axis into the page. For the variable u, we have taken unit vectors $u = (u^1, u^2) = (\cos\varphi, \sin\varphi)$, for $0 \leq \varphi < 2\pi$. The horizontal axis is the φ axis. At the equator ($\theta = 0$), L has a constant value of 1, since infinitesimal distances in all directions are the same there. $L = 1$ also when $\varphi = \pi/2$ because distances in the northern direction are the same for all θ and are like distances at the equator. When $\varphi = 0$ or π, the heading is due east or due west. As θ increases or decreases from 0, each degree of longitude represents a smaller and smaller distance

Figure 4.9 presents the action function L. Since L is independent of λ, we display its dependence on θ only. Regarding its dependence on direction, we use $u = (u^1, u^2) = (\cos\varphi, \sin\varphi)$, for $0 \leq \varphi < 2\pi$.

With the action function (4.3) in hand, we can now compute the length of the worldline $x(\sigma) = (\lambda(\sigma), \theta(\sigma))$, $0 \leq \sigma \leq 1$, with $P = (\lambda(0), \theta(0))$ and $Q = (\lambda(1), \theta(1))$. Recall our partition of $[0, 1]$ into n subintervals, each of length $\epsilon = 1/n$, and that $x_m = (\lambda(m\epsilon), \theta(m\epsilon))$. From the definition of the derivative, for any m, the displacement between two points $x_{m+1} = (\lambda((m+1)\epsilon), \theta((m+1)\epsilon))$ and $x_m = (\lambda(m\epsilon), \theta(m\epsilon))$ is approximately $(\Delta\lambda, \Delta\theta) = \epsilon\left(\frac{d\lambda}{d\sigma}(m\epsilon), \frac{d\theta}{d\sigma}(m\epsilon)\right)$. For ease of notation, denote differentiation by σ with a prime. If we define $u^1 = \frac{d\lambda}{d\sigma} = \lambda'$ and $u^2 = \theta'$, the length of the path $x(\sigma)$ is approximately

$$S_n(x) = \sum_{m=0}^{n-1} L\left(x(m\epsilon), x'(m\epsilon)\right)\epsilon.$$

The sum $S_n(x)$ is called a *Riemann sum*. As $n \to \infty$, S_n converges to

$$S(x) = \int_0^1 L\left(x(\sigma), x'(\sigma)\right)\, d\sigma, \tag{4.4}$$

where $L(x, u)$ is defined by (4.3).

Consider now motion along a route on the globe, as a function of the time t. On the map, the route is represented by a curve $x(t) = (\lambda(t), \theta(t))$. As shown in Sect. 2.3, the path $x(t)$ is stationary if and only if it satisfies the *Euler-Lagrange equations* (2.32). We denote the derivative by t with a prime. First, we compute the momenta defined by (2.31):

$$p = \frac{R}{\sqrt{(\theta')^2 + \cos^2\theta(\lambda')^2}}(\lambda' \cos^2\theta, \theta'). \tag{4.5}$$

The force is

$$\frac{\partial L}{\partial x} = \left(0, -\frac{R(\lambda')^2 \sin\theta\cos\theta}{\sqrt{(\theta')^2 + \cos^2\theta(\lambda')^2}}\right). \tag{4.6}$$

To simplify the Euler-Lagrange equations, we chose a new parameter τ such that

$$\frac{d\tau}{dt} = \sqrt{(\theta')^2 + \cos^2\theta(\lambda')^2}$$

and use a dot to denote differentiation by τ. Note that with this parametrization, the denominator of (4.5) equals 1. Then, formula (4.5) becomes $p = R(\dot{\lambda}\cos^2\theta, \dot{\theta})$. Since L does not depend on λ, the Law of Conservation (2.33) implies that

$$\dot{\lambda}\cos^2\theta = a, \tag{4.7}$$

for some constant a. For θ, the Euler-Lagrange equation is $R\ddot{\theta} = -R\dot{\lambda}^2 \sin\theta\cos\theta$. Substituting $\dot{\lambda}$ from (4.7), we obtain

$$\ddot{\theta} = -a\dot{\lambda}\tan\theta. \tag{4.8}$$

Consider a path along a meridian of longitude: $\lambda = \lambda_0$. Then $\dot{\lambda} = 0$, and (4.7) is satisfied with $a = 0$. From (4.8), we now have $\ddot{\theta} = 0$, implying that $\theta = \alpha\tau + \beta$, for constants α and β. Thus, for any longitude λ_0, the longitudinal path $(\lambda(\tau), \theta(\tau)) = (\lambda_0, \alpha\tau + \beta)$ is stationary. Notice that a meridian of longitude is one half of a great circle.

Consider next a path $\theta = \theta_0$ along a parallel of latitude. From (4.7), it follows that $\dot{\lambda}$ is constant and nonzero in order that the path be non-trivial. It is clear that $\ddot{\theta} = 0$ in this case, and so Eq. (4.8) holds only when $\tan\theta = 0$, that is, when $\theta = 0$. This shows that the only stationary path along a parallel of latitude is along the equator. Note that the equator is a great circle. In fact, rotational symmetry implies that *geodesics on spheres are arcs of great circles.*

We provide an intuitive proof of this fact. In Fig. 4.10, all three circles go through the points $(0, 1)$ and $(0, -1)$. Between these two points, the shortest path of the three shown is the path which is closest to a straight line. This is the path that lies on the circle with the largest radius.

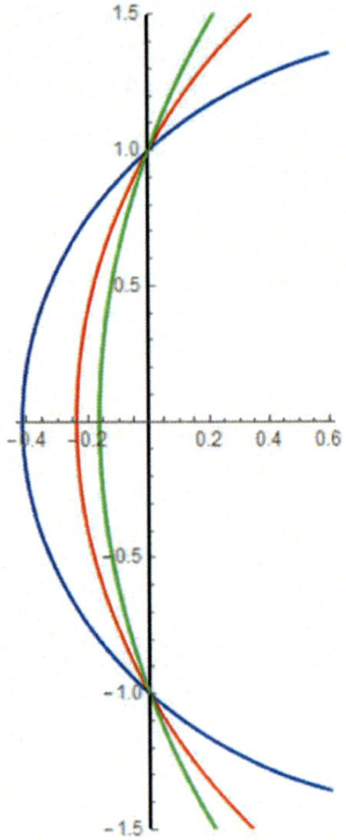

Fig. 4.10 Three arcs of circles all pass through the common points (0, 1) and (0, −1). The green arc is the shortest (closest to a straight line) among the three because its radius is the greatest

Equations of great circles may be derived as follows. Geometrically, a great circle on a sphere of radius R is the intersection of the sphere with a plane passing through the center of the sphere. Without loss of generality, we assume that the radius of the sphere is 1. From (4.1), a point on the unit globe with geographical coordinates λ, θ has Cartesian coordinates

$$x = \cos\lambda\cos\theta\ , \quad y = \sin\lambda\cos\theta\ , \quad z = \sin\theta. \tag{4.9}$$

The equation of a plane through the center of the sphere with normal vector (a, b, c) is $ax + by + cz = 0$. Thus, the equation of the corresponding great circle is

$$a\cos\lambda\cos\theta + b\sin\lambda\cos\theta + c\sin\theta = 0. \tag{4.10}$$

If $c = 0$, then either $\cos\theta = 0$, in which case (4.10) becomes trivial $(0 = 0)$, or $\cos\theta \neq 0$, and $\lambda = \arctan(-a/b)$, in which case (4.10) represents a great circle composed of two meridians separated by 180° of longitude. If $c \neq 0$, let $\tilde{a} = a/c$, $\tilde{b} = b/c$ and (4.10) becomes

$$\tilde{a}\cos\lambda\cos\theta + \tilde{b}\sin\lambda\cos\theta + \sin\theta = 0. \tag{4.11}$$

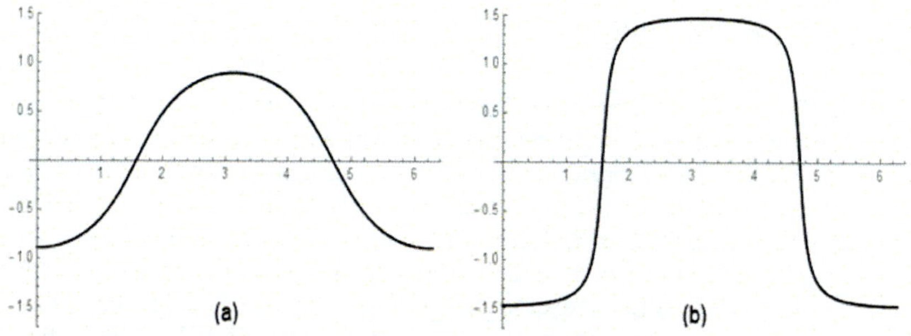

Fig. 4.11 Great circles on the Earth's surface, drawn on the flat map. The horizontal axis is longitude in radians. The vertical axis is latitude in radians. **a** The great circle with normal $(1, 0, 0.8)$. **b** The great circle with normal $(1, 0, 0.1)$

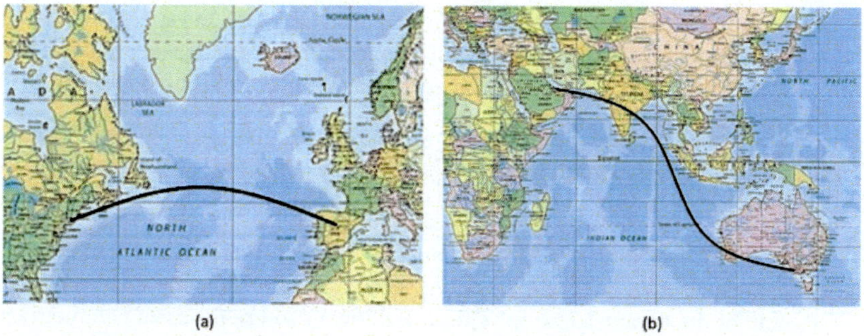

Fig. 4.12 **a** A geodesic from New York City ($74° W\ 40° N$) to Madrid ($3° W\ 40° N$). The two cities have the same latitude but the geodesic between them does not lie along a line of constant latitude. In fact, the geodesic heading from New York is $\approx 24.6°$ north of due east. **b** A geodesic from Dubai ($55°17'50'' E\ 25°15'47'' N$) to Melbourne ($144°58' E\ 37°49' S$). There is an inflection point at the equator

If $\cos\theta = 0$, then (4.11) implies that also $\sin\theta = 0$, which is a contradiction. If $\cos\theta \neq 0$, then dividing (4.11) by $\cos\theta$ and solving for θ leads to

$$\theta(\lambda) = \arctan\left(-\tilde{a}\cos\lambda - \tilde{b}\sin\lambda\right). \tag{4.12}$$

The equatorial plane, for example, has normal $(a, b, c) = (0, 0, c)$, and so (4.12) becomes $\theta(\lambda) \equiv 0$. See Figs. 4.11 and 4.12.

4.3 The Geometric Action Function and Its Properties

In this section, we apply the ideas of the previous sections and derive the properties that a distance or action function must have in order to describe relativistic dynamics.

The first step is the decision to work in an *inertial* frame in flat spacetime and not in a non-inertial frame in the object's spacetime. All of the calculations will be carried out in the inertial frame. This simplifies the model significantly and is akin to calculating geographic distances on a flat map instead of on the globe itself.

In order to define the shortest worldline, we need to define the "length" of a worldline in an object's spacetime. To do this, it is enough to define the distance between two infinitesimally close points P and Q in spacetime. This is the analog of the line element of the spacetime metric in GR [59, 78]. We propose the following definition.

Definition 4.1 The action function $L(x,u)$ is a scalar-valued function of a spacetime position x and a four-vector u, with the meaning that the distance between two points $P=x$ and $Q=x+\epsilon u$ in an object's spacetime is $L(x,u)\epsilon$ if ϵ is small.

Mathematically, the action function $L(x,u)=\lim_{\epsilon\to 0}\frac{D(x,x+u\epsilon)}{\epsilon}$, where $D(x,y)$ denotes the distance in the object's spacetime between two spacetime points x and y.

From definition (2.29), the length of a worldline ostensibly depends on the parametrization σ of the worldline. However, we want the action to be independent of the parametrization. This will allow us to use different parametrizations for different problems, since a judicious choice of parameter can simplify the calculations. Therefore, we obtain now a sufficient condition for the action function $L(x,u)$ to be independent of the choice of parameter.

Let σ' be another parametrization of the same worldline. Define a function $f(\sigma)=\sigma'$ such that $x(\sigma)=x(\sigma')$. Since the parametrizations must preserve order, $f'(\sigma)>0$. Now $\frac{dx}{d\sigma}=f'(\sigma)\frac{dx}{d\sigma'}$ and $d\sigma=\frac{1}{f'(\sigma)}d\sigma'$, implying that

$$L\left(x(\sigma),\frac{dx}{d\sigma}\right)d\sigma=L\left(x(\sigma'),f'(\sigma)\frac{dx}{d\sigma'}\right)\frac{1}{f'(\sigma)}d\sigma'.$$

Thus, in order to have $L\left(x(\sigma),\frac{dx}{d\sigma}\right)d\sigma=L\left(x(\sigma'),\frac{dx}{d\sigma'}\right)d\sigma'$, we need

$$L\left(x(\sigma'),f'(\sigma)\frac{dx}{d\sigma'}\right)=f'(\sigma)L\left(x(\sigma'),\frac{dx}{d\sigma'}\right). \tag{4.13}$$

In other words, for any positive scalar a, we require that $L(x,au)=aL(x,u)$. This means that $L(x,u)$ must be *positive homogeneous in u of degree* 1.

In light of Definition 4.1, the Principle of Relativity, and the above considerations, the action function $L(x,u)$ must satisfy the following properties:

1. $L(x, u)$ must be scalar valued and Lorentz invariant in order to satisfy the Principle of Relativity.
2. For the distance (action) $S[x(\sigma)] = \int L(x(\sigma), u(\sigma))d\sigma$ to be independent of the parametrization σ, $L(x, u)$ must be positive homogeneous in u of degree 1.
3. When the strength of the field goes to zero, we must have $L(x, u) = \sqrt{\eta_{\mu\nu}u^\mu u^\nu}$, the action function in Minkowski space.
4. Since the acceleration of a charge in an electromagnetic field depends only on the charge-to-mass ratio q/m, and the motion of a photon in a medium depends only on its frequency, $L(x, u)$ must depend on at most one parameter intrinsic to the object.

In the next section, we construct an action function with all of these properties.

4.4 Simple Action Function

As simplicity is one of the main themes of this book, we seek the *simplest* function $L(x, u)$ with the above properties. Recall Occam's razor: "explanations that posit fewer entities, or fewer kinds of entities, are to be preferred to explanations that posit more." And Einstein: "a physical theory should be as simple as possible, but not simpler."

First, we take care of Lorentz invariance (property 1). A simple way to construct a Lorentz-invariant scalar-valued function of a four-vector u is to contract m copies of u^μ with a tensor with m lower indices. For $m = 1$, such a function has the form $a_\mu(x)u^\mu$. For $m = 2$, we have $g_{\mu\nu}(x)u^\mu u^\nu$. For $m = 3$, the form of the function is $b_{\mu\nu\eta}(x)u^\mu u^\nu u^\eta$, and so on. Note that any scalar-valued function of these basic Lorentz-invariant scalar-valued functions is also Lorentz invariant.

Now for property 2. Note, for example, that $g_{\mu\nu}(x)(au)^\mu(au)^\nu = a^2 g_{\mu\nu}(x)u^\mu u^\nu$ is positive homogenous of degree 2. Nevertheless, $\sqrt{g_{\mu\nu}(x)u^\mu u^\nu}$ is positive homogenous of degree 1. Therefore, any linear combination of $a_\mu(x)u^\mu$, $\sqrt{g_{\mu\nu}(x)u^\mu u^\nu}$, $\sqrt[3]{b_{\mu\nu\eta}(x)u^\mu u^\nu u^\eta}$, and so on is positive homogeneous in u of degree 1 as well as Lorentz invariant.

To fulfill property 3, $L(x, u)$ must contain a term of the form $\sqrt{g_{\mu\nu}(x)u^\mu u^\nu}$ of order $m = 2$. We may assume that $g_{\mu\nu}$ is symmetric and thus depends on ten parameters. (If $g_{\mu\nu}$ is not symmetric, replace it with $h_{\mu\nu} = (g_{\mu\nu} + g_{\nu\mu})/2$.) The action $L(x, u) = \sqrt{g_{\mu\nu}(x)u^\mu u^\nu}$ is the action function in GR. Using property 3, we will write this term as $\sqrt{(\eta_{\mu\nu} + h_{\mu\nu}(x))u^\mu u^\nu}$, where $h_{\mu\nu}(x)$ is a symmetric tensor tending to zero when the strength of the field goes to zero.

At this point, we propose a further simplification and assume that there exists a four-covector-valued function $l_\mu(x)$ such that $h_{\mu\nu}(x) = -l_\mu(x)l_\nu(x)$. Thus,

$$g_{\mu\nu}(x) = \eta_{\mu\nu} - l_\mu(x)l_\nu(x). \tag{4.14}$$

This term in the action then becomes $\sqrt{\eta_{\mu\nu}u^\mu u^\nu - (l_\mu(x)u^\mu)^2}$. The minus sign is needed to avoid superluminal motion. Since the influenced direction of a gravitational field at each spacetime point is represented by a single null covector, for such a field we assume that $l_\mu(x)$ is a null covector in the direction of the propagation of the field. This reduces the ten free parameters of $g_{\mu\nu}(x)$ to the three free parameters of $l_\mu(x)$. This simplification of a gravitational field description is justified because the covector l_μ exists for a single-source gravitational field in all of the following cases (see [63, 64], and [2, Chap. 7]):

- any static field;
- the field of a spherically symmetric, non-rotating body;
- the field of a rotating black hole.

Our description of the gravitational field for a single point, static, spherically symmetric source is similar to Whitehead's theory of gravitation [96]. It was shown in [24] and [31] that for this case, Whitehead's theory is equivalent to the standard Schwarzschild metric. For a field generated by several sources, however, Whitehead's theory differs from ours. Whitehead assumes *linear* dependence of the total four-potential $l_\mu(x)$ of the field on the four-potentials of the individual sources. In other words, for Whitehead, $l_\mu(\mathbf{x}) = \sum_k l_\mu^{(k)}(\mathbf{x})$. Based on this linearity assumption, Whitehead's theory predicts several new predictions which were shown in [53] to contradict experimental observations. Our model, on the other hand, does *not* assume additivity of the four-potentials. In fact, in section 6.7, we will show that in our model, the square of l_μ is the sum of the squares of the $l^{(k)}$ s. Thus, these experiments do not present any difficulty for our approach.

From the Definition 4.1 of the action function $L(x, u)$, the dependence on u is infinitesimal. This indicates that we need low-order approximations to define it properly. Thus, for the sake of simplicity, we assume that the action function consists of terms of order 1 and 2 only. This leads us to the following simple action function for an object's spacetime:

$$L(x, u) = \sqrt{\eta_{\mu\nu}u^\mu u^\nu - (l_\mu(x)u^\mu)^2} + kA_\mu(x)u^\mu. \tag{4.15}$$

The spacetime under consideration may be influenced by fields or isotropic media. By a *field*, we mean that there is a source, such as a massive object and/or a charge/current distribution. The sources act at a distance, and their influence propagates with the speed of light. For media, there is no source. Nevertheless, the medium may be at rest or in motion.

If the spacetime is influenced by fields, then as long as l_μ and A_μ are four-covector-valued functions of one of the forms (3.83) or (3.84), the action function $L(x, u)$ of (4.15) satisfies properties 1 and 2. If the spacetime is influenced by an isotropic medium, then $l = (l_0, 0, 0, 0)$ when the medium is at rest. For moving media, l is computed by Lorentz transforming this covector from the frame comoving with the medium. The covector l_μ goes to zero when the field strength goes to zero (by construction), so if the same is true of A_μ, then $L(x, u)$ will satisfy property 3. As for property 4, the function $l_\mu(x)$ defines the gravitational field, and in this case l is a null covector in the direction of the field propagation. It does not depend on the moving object at all. Note that $L(x, u)$ is not affected if we replace l by $-l$. This reflects the fact that the gravitational force is only attractive. As we will see in the next section, the function $A_\mu(x)$ is the four-potential of the electromagnetic field. For a charge moving in this field, we will have $k \sim q/m$, and thus $L(x, u)$ depends only on the charge-to-mass ratio. Replacing k by $-k$ will change the action, reflecting the fact that the electric force is both attractive and repulsive. For the motion of a photon in isotropic medium, $l_\mu(x)$ depends on the properties of the medium and the frequency of the photon.

Since $L(x, u)$ must be real valued, the expression under the square root in (4.15) must be non-negative. Since we will substitute four-velocities for u, this limitation implies that there is a domain of admissible four-velocities

$$D_x(l) = \{u : \eta_{\mu\nu}u^\mu u^\nu - (l_\mu(x)u^\mu)^2 \geq 0\}. \tag{4.16}$$

We assume that the four-velocity of a massive object is always in the interior of $D_x(l)$ and that the four-velocity of a massless particle belongs to the boundary $\partial D_x(l)$ of $D_x(l)$. If $l = 0$, then $D_x(l)$ is the same as in SR.

Mathematically, our action function plays a similar role to that of the classical Lagrangian $L = T - U$ (kinetic minus potential energy). However, the similarity ends there. The classical Lagrangian depends on the mass of the object in motion, and yet it was already known experimentally and logically by Galileo Galilei that motion in a gravitational field is independent of mass. In other words, the classical Lagrangian contains a parameter which has no bearing on the physics! On the other hand, our action function for a gravitational field does not depend on the mass of the object in motion. It is object independent, as in GR and as it should be. For an electromagnetic field, our action function is object dependent and depends on the charge-to-mass ratio. This is also as it should be, since the motion in this case does indeed depend on this ratio. We also point out that the physical significance of the quantity $T - U$ of the classical Lagrangian is not clear. On the other hand, the physical meaning of our action function is clear—it describes distances in spacetime.

4.5 Universal Relativistic Equation of Motion

In this section, we apply the Euler-Lagrange equations (2.32) to the action function (4.15). We assume throughout that l is a null vector. This leads to a relativistic dynamics equation of motion which encompasses both electromagnetism and gravity. Motion in isotropic media will be handled separately, in Chap. 7, where l will no longer be a null vector.

The unit-free energy-momentum covector is

$$p_\lambda = \frac{\partial L(x,u)}{\partial u^\lambda}\bigg|_{x=x(\sigma),\, u=dx(\sigma)/d\sigma} = \frac{\eta_{\lambda\mu}\frac{dx^\mu}{d\sigma} - l_\mu \frac{dx^\mu}{d\sigma} l_\lambda}{\sqrt{\eta_{\mu\nu}\frac{dx^\mu}{d\sigma}\frac{dx^\nu}{d\sigma} - \left(l_\mu \frac{dx^\mu}{d\sigma}\right)^2}} + kA_\lambda. \tag{4.17}$$

Now we come to the choice of the parameter σ. Two obvious candidates present themselves—the *proper time* τ of the object in motion, defined by

$$d\tau^2 = \eta_{\mu\nu} dx^\mu dx^\nu, \tag{4.18}$$

and the parameter $\tilde{\tau}$, defined by

$$d\tilde{\tau}^2 = \eta_{\mu\nu} dx^\mu dx^\nu - (l_\mu dx^\mu)^2. \tag{4.19}$$

We now explore the physical meaning of these two parameters and derive two relativistic dynamics equations, one for τ and one for $\tilde{\tau}$. We begin with proper time.

4.5.1 The Equation of Motion Using Proper Time

The proper time is a natural choice since it is a Lorentz-invariant scalar and, thus, the same for all inertial observers. In addition, proper time does not depend on the influenced spacetime. Recall from Sect. 3.4 that one may think of proper time as the time displayed by a clock moving with the object. For a clock at rest in the inertial lab frame, we have $cdt = d\tau$. In general, for a clock with velocity $\mathbf{v}$ in K, we have

$$cdt = \gamma d\tau, \tag{4.20}$$

where $\gamma = \gamma(\mathbf{v})$ is defined by (3.19). Here, differentiation by τ will be denoted by a dot. Thus, the four-velocity is $\dot{x} = dx(\tau)/d\tau$ and $\eta_{\mu\nu}\dot{x}^\mu\dot{x}^\nu = 1$.

Using proper time, the energy-momenta (4.17) are

$$p_\lambda = \frac{\dot{x}_\lambda - (l \cdot \dot{x}) l_\lambda}{\sqrt{1 - (l \cdot \dot{x})^2}} + kA_\lambda. \tag{4.21}$$

When $l = A = 0$, we are in Minkowski space, and p_λ reduces to the four- velocity $\dot{x}_\lambda$. As we will soon show explicitly, the term $\sqrt{1-(l\cdot\dot{x})^2}$ is an expression of the *gravitational time dilation*. For the moment, we remark only that this term does lead to lengthy expressions for the derivatives of p_λ. Nevertheless, the resulting equation of motion is not so complicated, as many terms cancel out along the way.

To apply the Euler-Lagrange equations (2.32), we first calculate the derivative of the unit-free energy-momentum covector. To differentiate a function of x by τ, we use the chain rule. For example,

$$\frac{d}{d\tau}A_\lambda(x) = \frac{dA_\lambda}{dx^\nu}\frac{dx^\nu}{d\tau} = A_{\lambda,\nu}\dot{x}^\nu. \tag{4.22}$$

Case 1 $l = 0$

From (4.21), the momenta are

$$p_\lambda = \dot{x}_\lambda + kA_\lambda. \tag{4.23}$$

The momentum is the sum of two terms. The first term is the particle's momentum p_λ^{fr} as a free particle (see Eq. (3.60)). The second term is the momentum imparted to the particle by the field generated by A.

The τ derivative of the momentum is $\dot{p}_\lambda = \ddot{x}_\lambda + kA_{\lambda,\nu}\dot{x}^\nu$, and the four-force is $\frac{\partial L}{\partial x^\lambda} = kA_{\nu,\lambda}\dot{x}^\nu$. Thus, the Euler-Lagrange equations (2.32) become

$$\ddot{x}_\lambda = kA_{\nu,\lambda}\dot{x}^\nu - kA_{\lambda,\nu}\dot{x}^\nu. \tag{4.24}$$

In Newtonian dynamics, the force is the time derivative of the momentum (Eq. (2.1)). But there, the momentum is purely the momentum that the object has as a result of its mass and velocity, its momentum as a free particle. Thus, Newton's Second Law says that the force on an object equals the time derivative of its momentum as a free particle. There is no notion, however, of field momentum. In relativistic dynamics, the situation is different. The force on an object is still the derivative of its momentum, but now the momentum is the sum of its momentum as a free particle and its field momentum. Hence, to obtain the object's *acceleration*, which is the derivative of the free momentum, one must *subtract* the change $kA_{\lambda,\nu}(x)\dot{x}^\nu$ in the field momentum from the acceleration $kA_{\nu,\lambda}(x)\dot{x}^\nu$ due to the force. This is seen explicitly is Eq. (4.24).

To simplify the equation of motion (4.24), we introduce a first-order derivative F of a covector-valued function $f(x)$ by

$$F_{\lambda\nu}(f(x)) = f_{\nu,\lambda} - f_{\lambda,\nu}. \tag{4.25}$$

F is a rank 2 antisymmetric tensor. Raising one of the indices, we have

$$F_\nu^\alpha(f(x)) = \eta^{\alpha\lambda}F_{\lambda\nu}(f(x)). \tag{4.26}$$

In this notation, we can rewrite (4.24) as

$$\ddot{x}^\alpha = kF_\nu^\alpha(A)\dot{x}^\nu. \tag{4.27}$$

This is the relativistic equation of motion in a field defined by the *linear four-potential* $A(x)$. For

$$k = \frac{q}{mc^2}, \tag{4.28}$$

equation (4.27) is the equation of motion of a particle with charge-to-mass ratio q/m in an electromagnetic field with field strength tensor $F^{\alpha}_{\nu}(A)$ (see [61, Eq. 12.3]).

By a *solution* of Eq. (4.27), we mean a function $f(\tau)$ such that

$$\dot{f}^{\alpha} = kF^{\alpha}_{\nu}(A)f^{\nu}. \tag{4.29}$$

We prove the following claim.

Claim Let $f(\tau)$ and $g(\tau)$ be solutions to (4.27). Then

(a) $\dot{f} \cdot f = 0$
(b) $f(\tau) \cdot g(\tau)$ is constant.

Note that when $f(\tau)$ is a four-velocity, then part (a) of the claim says that the four-acceleration is orthogonal to the four-velocity, as it is required to be (see the discussion after (3.43)).

To prove (a), we use (4.29) and the antisymmetry of F:

$$\dot{f} \cdot f = \dot{f}_{\mu} f^{\mu} = kF_{\mu\nu} f^{\nu} f^{\mu} = kF_{\nu\mu} f^{\nu} f^{\mu} = -kF_{\mu\nu} f^{\nu} f^{\mu}.$$

The third equality holds by symmetry of the dummy indices. The fourth equality holds by the antisymmetry of $F_{\mu\nu}$. Finally, since $F_{\mu\nu} f^{\nu} f^{\mu} = -F_{\mu\nu} f^{\nu} f^{\mu}$, this must be 0.

For (b), differentiate:

$$\frac{d}{d\tau}(f \cdot g) = \frac{d}{d\tau}(\eta_{\mu\nu} f^{\mu} g^{\nu}) = \eta_{\mu\nu}(\dot{f}^{\mu} g^{\nu} + f^{\mu} \dot{g}^{\nu})$$

$$= \eta_{\mu\nu}\left(kF^{\mu}_{\lambda} f^{\lambda} g^{\nu} + kf^{\mu} F^{\nu}_{\lambda} g^{\lambda}\right) =$$

$$= k\left(F_{\nu\lambda} f^{\lambda} g^{\nu} + f^{\mu} F_{\mu\lambda} g^{\lambda}\right) = k\left(F_{\nu\mu} + F_{\mu\nu}\right) f^{\mu} g^{\nu}.$$

Since F is antisymmetric, we have $\frac{d}{d\tau}(f \cdot g) = 0$ and hence, $f(\tau) \cdot g(\tau)$ is constant. In particular, the norm $f^2(\tau)$ of any solution is conserved along the worldline. This implies that any solution $x(\tau)$ of (4.27) with initial $(\dot{x})^2(0) = 1$ will satisfy $(\dot{x})^2(\tau) = 1$ for all τ. This proves the claim.

In fact, it is known (see [18, p. 1–65]) that for a given initial condition $\dot{x}(0)$, (4.27) has the unique solution

$$\dot{x}(\tau) = \exp(kF(A)\tau)\dot{x}(0) = \left(\sum_{n=0}^{\infty} k^n \tau^n F(A)^n\right) \dot{x}(0). \tag{4.30}$$

Case 2 $A = 0$

From (4.21), the momenta are

$$p_\lambda = \frac{\dot{x}_\lambda - (l \cdot \dot{x}) l_\lambda}{\sqrt{1 - (l \cdot \dot{x})^2}}. \tag{4.31}$$

The τ derivative of the momentum is

$$\dot{p}_\lambda = \frac{\ddot{x}_\lambda - (l \cdot \ddot{x} + \dot{l} \cdot \dot{x}) l_\lambda - (l \cdot \dot{x}) l_{\lambda,\nu} \dot{x}^\nu}{(1 - (l \cdot \dot{x})^2)^{-1/2}} +$$

$$+ \frac{(\dot{x}_\lambda - (l \cdot \dot{x}) l_\lambda)(l \cdot \dot{x})(l \cdot \ddot{x} + \dot{l} \cdot \dot{x})}{(1 - (l \cdot \dot{x})^2)^{-3/2}}.$$

Let

$$d_\lambda = \frac{l_\lambda - (l \cdot \dot{x}) \dot{x}_\lambda}{1 - (l \cdot \dot{x})^2}.$$

Opening the parentheses and cancelling, we obtain

$$\dot{p}_\lambda = \frac{1}{\sqrt{1 - (l \cdot \dot{x})^2}} \left(\ddot{x}_\lambda - (l \cdot \ddot{x}) d_\lambda - (\dot{l} \cdot \dot{x}) d_\lambda - (l \cdot \dot{x}) l_{\lambda,\nu} \dot{x}^\nu \right). \tag{4.32}$$

Note that $\dot{l} \cdot \dot{x}$ can be written as

$$\dot{l} \cdot \dot{x} = l_{\mu,\nu} \dot{x}^\mu \dot{x}^\nu = J_{\mu\nu}(l) \dot{x}^\mu \dot{x}^\nu, \tag{4.33}$$

where $J_{\mu\nu}(l)$ is the Jacobian matrix of l. By symmetry, we have

$$\dot{l} \cdot \dot{x} = J^s_{\mu\nu}(l) \dot{x}^\mu \dot{x}^\nu, \tag{4.34}$$

where $J^s_{\mu\nu}$ is the symmetric part of $J_{\mu\nu}$.

The four-force in this case is

$$\frac{\partial L}{\partial x^\lambda} = -\frac{(l \cdot \dot{x}) l_{\nu,\lambda} \dot{x}^\nu}{\sqrt{1 - (l \cdot \dot{x})^2}}. \tag{4.35}$$

Thus, the Euler-Lagrange equations (2.32) become

$$\ddot{x}_\lambda - (l \cdot \ddot{x}) d_\lambda - (\dot{l} \cdot \dot{x}) d_\lambda = -(l \cdot \dot{x})(l_{\nu,\lambda} - l_{\lambda,\nu}) \dot{x}^\nu. \tag{4.36}$$

Using (4.25), this equation becomes

$$\ddot{x}_\lambda - (l \cdot \ddot{x}) d_\lambda - (\dot{l} \cdot \dot{x}) d_\lambda = -(l \cdot \dot{x}) F_{\lambda\nu}(l) \dot{x}^\nu. \tag{4.37}$$

To obtain an explicit formula for the acceleration, take the inner product of l with (4.37):

$$(l \cdot \ddot{x})(1 - d \cdot l) - (\dot{l} \cdot \dot{x})(d \cdot l) = -(l \cdot \dot{x}) F_{\lambda\nu}(l) \dot{x}^\nu l^\lambda.$$

Since l is a null vector, we have $1 - d \cdot l = (1 - (l \cdot \dot{x})^2)^{-1}$ and $(1 - (l \cdot \dot{x})^2) d \cdot l = -(l \cdot \dot{x})^2$, so the above formula yields

$$l \cdot \ddot{x} = -(\dot{l} \cdot \dot{x})(l \cdot \dot{x})^2 - (l \cdot \dot{x})(1 - (l \cdot \dot{x})^2) F_{\lambda\nu}(l) \dot{x}^\nu l^\lambda.$$

Substituting this into (4.37), denoting

$$l^\perp = l - (l \cdot \dot{x}) \dot{x}, \tag{4.38}$$

and raising the index, we obtain an explicit formula for $\ddot{x}^\alpha$:

$$\ddot{x}^\alpha = -(l \cdot \dot{x}) F^\alpha_\nu(l) \dot{x}^\nu - (l \cdot \dot{x}) F^\lambda_\nu(l) \dot{x}^\nu l_\lambda (l^\perp)^\alpha + (\dot{l} \cdot \dot{x})(l^\perp)^\alpha. \tag{4.39}$$

To simplify the notation, we define

$$b^\alpha(l) = -(l \cdot \dot{x}) F^\alpha_\nu(l) \dot{x}^\nu \tag{4.40}$$

Using this and (4.33), we can rewrite Eq. (4.39) as

$$\ddot{x} = b + (b \cdot l) l^\perp + l_{\mu,\nu} \dot{x}^\mu \dot{x}^\nu l^\perp. \tag{4.41}$$

This is the equation for motion of objects in a gravitational field defined by the *quadratic four-potential* $l(x)$.

Before moving on to case 3 ($A \neq 0, l \neq 0$), it is worthwhile to analyze Eq. (4.41). The first term, b, can be written as a multiple of the antisymmetric part $J^a_{\mu\nu}(l)$ of the Jacobian matrix of l. Thus, this term is akin to the lone term on the right-hand side of (4.27). Indeed, in the Newtonian limit of a static gravitational field, this term becomes the classical acceleration due to an inverse-square, Coulomb-type field. In the same limit, the second term, $(b \cdot l) l^\perp$, corrects the first term for the gravitational time dilation. See Sect. 6.7 for details. The third term, in this limit, vanishes, since the field is static.

One should check that the four-acceleration is orthogonal to the four-velocity. Here, in fact, we have even more. Each of the three components of $\ddot{x}$ is orthogonal to $\dot{x}$. By part (a) of the above claim, b is orthogonal to the four-velocity. The second and third terms of the acceleration are parallel to $l^\perp$, which is orthogonal to $\dot{x}$ by definition. It is these two terms that ensure that an object's four-velocity remains within the admissible region.

To see this, consider an object whose four-velocity $\dot{x}$ lies on the boundary of admissible four-velocities. Then the expression under the square root in the action function (4.15) vanishes, implying that $(l \cdot \dot{x})^2 = 1$. Differentiating this by τ yields

$$\dot{l} \cdot \dot{x} + l \cdot \ddot{x} = 0. \tag{4.42}$$

Now, suppose that $\ddot{x} = b + y$. Using this and (4.42), we have

$$-\dot{l} \cdot \dot{x} = l \cdot \ddot{x} = l \cdot b + l \cdot y.$$

This implies that

$$l \cdot y = -(\dot{l} \cdot \dot{x} + l \cdot b). \tag{4.43}$$

We check that $y = ((b \cdot l) + (\dot{l} \cdot \dot{x}))l^{\perp}$ satisfies (4.43). Using $(l \cdot \dot{x})^2 = 1$, the definition of $l^{\perp}$ and the fact that l is null, we have

$$l \cdot ((b \cdot l) + (\dot{l} \cdot \dot{x}))l^{\perp} = -(l \cdot \dot{x})^2((b \cdot l) + (\dot{l} \cdot \dot{x})) = -(b \cdot l + \dot{l} \cdot \dot{x}).$$

The first two terms of the equation of motion (4.41), taken together, are perpendicular on the boundary to $l^{\perp}$ and thus, to the third term. To show this, we recall that $l \cdot \dot{x} = 1$ on the boundary ∂D_x and rely on the following lemma, whose proof we leave as an exercise.

Lemma

(a) $b \cdot l = b \cdot l^{\perp}$
(b) On the boundary ∂D_x, $(l^{\perp})^2 = -1$.

Using the lemma, we have

$$l^{\perp} \cdot (b + (b \cdot l)l^{\perp}) = l^{\perp} \cdot b + (b \cdot l)(l^{\perp})^2 = l^{\perp} \cdot b + (b \cdot l^{\perp})(l^{\perp})^2$$
$$= (1 + (l^{\perp})^2)(b \cdot l^{\perp}) = 0.$$

Case 3 $A \neq 0, l \neq 0$

The momenta (4.21) may be decomposed by the *source* of the momentum. Thus, using G to signify gravity and EM to signify electromagnetism, we may write

$$p_\lambda = p_\lambda^{\mathrm{fr}} + p_\lambda^{\mathrm{G}} + p_\lambda^{\mathrm{EM}}, \tag{4.44}$$

where

$$p_\lambda^{\mathrm{fr}} = \dot{x}_\lambda \ , \quad p_\lambda^{\mathrm{G}} = \frac{\dot{x}_\lambda - (l \cdot \dot{x})l_\lambda}{\sqrt{1 - (l \cdot \dot{x})^2}} - \dot{x}_\lambda \ , \quad p_\lambda^{\mathrm{EM}} = kA_\lambda. \tag{4.45}$$

The quantity p_λ^{fr} is the unit-free energy-momentum of a free particle with four-velocity $\dot{x}$ (see (3.60)). If multiplied by mc, it becomes the standard relativistic four-momentum (3.61). The terms p_λ^{G} and p_λ^{EM} are the additional energy-momenta imparted to the particle by the gravitational and electromagnetic fields, respectively. We call these the particle's *field energy-momenta*.

To derive the equation of motion in this general case, we modify the above equations of Case 2, as follows. In Eq. (4.32) for $\dot{p}_\lambda$, we have to add an additional term $kA_{\lambda,\nu}\dot{x}^\nu$. This leads to replacing the antisymmetric tensor $F_{\lambda\nu}(l)$ in Eq. (4.37) with a new antisymmetric tensor $\tilde{F}$, defined by

$$\tilde{F}_{\lambda\nu}(A,l) = \sqrt{1-(l\cdot\dot{x})^2}kF_{\lambda\nu}(A) - (l\cdot\dot{x})F_{\lambda\nu}(l). \tag{4.46}$$

The equation of motion is then

$$\ddot{x}^\alpha = \tilde{F}^\alpha_\nu(A,l)\dot{x}^\nu + \tilde{F}^\lambda_\nu(A,l)\dot{x}^\nu l_\lambda (l^\perp)^\alpha + (\dot{l}\cdot\dot{x})(l^\perp)^\alpha. \tag{4.47}$$

Defining

$$\tilde{b}^\alpha(A,l) = \tilde{F}^\alpha_\nu(A,l)\dot{x}^\nu, \tag{4.48}$$

the equation of motion in a field defined by $A(x)$ and $l(x)$ is

$$\ddot{x} = \tilde{b} + (\tilde{b}\cdot l)l^\perp + l_{\mu,\nu}\dot{x}^\mu\dot{x}^\nu l^\perp. \tag{4.49}$$

The same argument as for case 2 shows that the four-velocity of an object remains admissible throughout the evolution and that the sum of the first two terms is orthogonal to $l^\perp$. Here, one relies on the fact that, since F is antisymmetric, so is $\tilde{F}$.

In practice, we usually have no need to actually *solve* the dynamics equation (4.49). Instead, we use conservations guaranteed by the Euler-Lagrange equations. Typical examples are conservation of energy and conservation of angular momentum. These reduce the above second-order differential equations to first order.

For a physical system with not enough conserved quantities, the above second-order equations may be solved numerically using the relationships

$$x^\alpha(\tau+\Delta\tau) = x^\alpha(\tau) + \dot{x}^\alpha\Delta\tau$$
$$\dot{x}^\alpha(\tau+\Delta\tau) = \dot{x}^\alpha(\tau) + \ddot{x}^\alpha\Delta\tau.$$

4.5.2 The Equation of Motion Using $\tilde{\tau}$

Here, we derive the equation of motion using the parameter $\tilde{\tau}$, defined by (4.19). Indeed, simplicity would seem to suggest using the parameter $\tilde{\tau}$ because it makes the denominator of (4.17) equal to 1. Thus, the expressions for p_λ and $dp_\lambda/d\tilde{\tau}$ are simpler than in the case of proper time. Indeed, as long as our field has a single source, we feel free to use $\tilde{\tau}$ instead of proper time. However, when there are multiple sources, each one with its own $\tilde{\tau}$, we must use proper time.

To understand the physical meaning of $\tilde{\tau}$, consider the case in which the object is at rest at a spacetime point x in K. This means that $dx^j = 0$, for $j = 1, 2, 3$. Substituting $dx^0 = cdt$ into (4.19), we have $d\tilde{\tau}^2 = c^2dt^2 - l_0^2(x)c^2dt^2 = c^2dt^2(1-l_0^2(x))$. Since gravity affects all objects (charged, non-charged, massive, and massless), it is natural to assume that clocks will be affected by gravity. This, in fact, was predicted by GR and verified experimentally by the Pound-Rebka [85] experiment, among others. We also make the assumption that, as in Minkowski space, if a particle travels with the speed of light in our inertial frame, then $d\tilde{\tau}^2 = 0$.

If we denote $d\tilde{\tau} = cdt'$ and interpret t' as the time of a clock at rest at x in the gravitational field, then the time t in the inertial lab frame is related to $\tilde{\tau}$ by

$$cdt = \frac{1}{\sqrt{1 - l_0^2(x)}} d\tilde{\tau} = \frac{1}{\sqrt{1 - l_0^2(x)}} cdt'. \tag{4.50}$$

For $l_0^2(r) = r_s/r$, where r_s is the Schwarzschild radius, this is the gravitational time dilation formula for a clock at rest in the gravitational field of a static, spherically symmetric body, as predicted by GR and verified experimentally.

Note that the A_μ term in the momentum (4.17) does not contribute to the parameter $\tilde{\tau}$. This is because, as we shall shortly see, A_μ is the four-potential of an electromagnetic field. Now an electrically neutral clock will not be affected by an electromagnetic field. Thus, the time dilation factor for a neutral clock is as in formula (4.50). Moreover, we do not *allow* charged clocks in our model. This would lead to the untenable situation in which even clocks which are infinitesimally close to each other have wildly different time dilation factors.

As noted above, using $\sigma = \tilde{\tau}$ makes the denominator of (4.17) equal to 1, and the unit-free energy-momenta are

$$p_\lambda = \eta_{\lambda\mu}\dot{x}^\mu - l_\mu(x)\dot{x}^\mu l_\lambda(x), \tag{4.51}$$

where here, the dot denotes differentiation by $\tilde{\tau}$. The $\tilde{\tau}$ derivative of the unit-free energy-momentum is

$$\dot{p}_\lambda = \eta_{\lambda\mu}\ddot{x}^\mu - l_\mu\ddot{x}^\mu l_\lambda - l_{\mu,\nu}\dot{x}^\nu\dot{x}^\mu l_\lambda - l_\mu\dot{x}^\mu l_{\lambda,\nu}\dot{x}^\nu + kA_{\lambda,\nu}\dot{x}^\nu. \tag{4.52}$$

The four-force is

$$\frac{\partial L(x,u)}{\partial x^\lambda} = -l_\mu\dot{x}^\mu l_{\nu,\lambda}\dot{x}^\nu + kA_{\nu,\lambda}\dot{x}^\nu,$$

and the Euler-Lagrange equations (2.32) imply that a stationary world line satisfies the relativistic dynamics equation

$$\eta_{\lambda\mu}\ddot{x}^\mu - l_\mu\ddot{x}^\mu l_\lambda = (l_{\lambda,\nu}l_\mu - l_{\nu,\lambda}l_\mu + l_{\mu,\nu}l_\lambda)\dot{x}^\mu\dot{x}^\nu + k(A_{\nu,\lambda} - A_{\lambda,\nu})\dot{x}^\nu, \tag{4.53}$$

or

$$\eta_{\lambda\mu}\ddot{x}^\mu - l_\mu\ddot{x}^\mu l_\lambda = (-F_{\lambda\nu}(l)l_\mu + l_{\mu,\nu}l_\lambda)\dot{x}^\mu\dot{x}^\nu + kF_{\lambda\nu}(A)\dot{x}^\nu, \tag{4.54}$$

with F defined by (4.25).

To obtain an explicit formula for the four-acceleration $\ddot{x}^\alpha$, we contract both sides of the previous equation with the tensor $T^{\alpha\lambda} = \eta^{\alpha\lambda} + l^\alpha l^\lambda$. Assuming that l is a null vector, the left-hand side of (4.54) becomes

$$\ddot{x}^{\alpha} - l_{\mu}\ddot{x}^{\mu}l^{\alpha} + l_{\mu}\ddot{x}^{\mu}l^{\alpha} - l_{\mu}\ddot{x}^{\mu}l_{\lambda}l^{\lambda}l^{\alpha} = \ddot{x}^{\alpha}\,. \tag{4.55}$$

Now apply $T^{\alpha\lambda}$ to the right-hand side of (4.54). Writing $l \cdot \dot{x}$ for $l_{\mu}\dot{x}^{\mu}$, the right-hand side becomes

$$-F^{\alpha}_{\nu}(l)\dot{x}^{\nu}(l\cdot\dot{x}) - F_{\lambda\nu}(l)\dot{x}^{\nu}l^{\lambda}(l\cdot\dot{x})l^{\alpha} + l_{\mu,\nu}\dot{x}^{\mu}\dot{x}^{\nu}l^{\alpha} + kF^{\alpha}_{\nu}(A)\dot{x}^{\nu} + kF_{\lambda\nu}(A)\dot{x}^{\nu}l^{\lambda}l^{\alpha}.$$

Combining this with (4.55) and denoting

$$\hat{b}^{\alpha} = (kF^{\alpha}_{\nu}(A) - (l\cdot\dot{x})F^{\alpha}_{\nu}(l))\dot{x}^{\nu},$$

Equation (4.54) becomes

$$\ddot{x}^{\alpha} = \hat{b}^{\alpha} + (\hat{b}\cdot l)l^{\alpha} + l_{\mu,\nu}\dot{x}^{\mu}\dot{x}^{\nu}l^{\alpha}\,. \tag{4.56}$$

Here we have used the fact that $F_{\lambda\nu}l^{\lambda} = F^{\lambda}_{\nu}l_{\lambda}$, whose proof we leave as an exercise.

Comparing the equations of motion with respect to τ (4.49) and $\tilde{\tau}$ (4.56), we notice that $\hat{b}$ differs from $\tilde{b}$ only in the gravitational time dilation factor near the electromagnetic tensor. The parameter $\tilde{\tau}$ incorporates this time dilation. Another difference is that the τ equation uses $l^{\perp}$, while the $\tilde{\tau}$ equation uses l.

Chapter 5
The Electromagnetic Field in Vacuum

This chapter is devoted to the electromagnetic force. We assume, throughout the chapter, that gravitational forces are negligible, being thirty-five orders weaker than electromagnetism. We derive the Liénard-Wiechert four-potential for a field generated by a moving charge and compute the resultant field. Our four-potential does not have a gauge and is thus uniquely defined. We compute explicitly the electric and magnetic fields of a uniformly moving source. We define a new notion of the *energy-momentum of a field*. The spatial part of the energy-momenta of a field of a source at rest vanishes. The zero component is the potential energy of the field. For a moving source, the energy-momentum of the source is transferred to the energy-momentum of the field. This explains the change in the direction of acceleration from the retarded direction. In addition, we interpret the appearance of the magnetic field as a result of the field's angular momentum with respect to a test charge. For an accelerating source, we obtain the far field. Next, we expand the model to incorporate multiple sources. Maxwell's equations are derived as well as a proof of the Biot-Savart Law. The later sections deal with orbits of particles in single-source electromagnetic fields.

5.1 The Electromagnetic Field Tensor

The simple action function $L(x, u)$, defined by (4.15), contains two four-covector-valued functions $l_\mu(x)$ and $A_\mu(x)$. In this chapter, we work with $A_\mu(x)$, explore its physical meaning, and show that it is associated with electromagnetic fields. Thus, we assume in this chapter that $l_\mu(x) = 0$, and our action function here is

$$L(x, u) = \sqrt{\eta_{\mu\nu} u^\mu u^\nu} + k A_\nu(x) u^\nu. \tag{5.1}$$

Y. Friedman and T. Scarr, *A Novel Approach to Relativistic Dynamics*,
Fundamental Theories of Physics 210,
https://doi.org/10.1007/978-3-031-25214-3_5

The parameter τ on the world line, defined by (3.38) and (4.18), is

$$d\tau^2 = \eta_{\mu\nu} dx^\mu dx^\nu. \tag{5.2}$$

Recall that τ has units of length. We denote differentiation by τ with a dot.

Let K be an inertial frame, with time t. The time on a clock at rest in K is $\tau = ct$. When the clock is moving, we have $\gamma d\tau = cdt$, where γ is the Lorentz factor (3.19) corresponding to the velocity of the clock in K.

In order to apply the Euler-Lagrange equations to the action (5.1), we need to compute both the energy-momenta $p_\mu(x)$ and the components $\partial L/\partial x^\mu$ of the four-force. From (4.45), the unit-free energy-momenta are

$$p_\lambda = p_\lambda^{\mathrm{fr}} + p_\lambda^{\mathrm{EM}} = \dot{x}_\lambda + kA_\lambda(x). \tag{5.3}$$

The first term on the right-hand side is the unit-free energy-momentum of a free particle with four-velocity $\dot{x}$ (see (3.60)). If multiplied by mc, it becomes the standard relativistic four-momentum (3.61). The term $kA_\lambda(x) = p_\lambda^{\mathrm{EM}}$ is the additional momentum imparted to the particle by the field. We call this the particle's *field momentum*.

As shown in Sect. 4.5, the Euler-Lagrange equations imply that the *equation of motion* is

$$\ddot{x}^\alpha = kF_\nu^\alpha(A)\dot{x}^\nu, \tag{5.4}$$

where $F_{\lambda\nu}(A) = A_{\nu,\lambda} - A_{\lambda,\nu}$ and $F_\nu^\alpha(A) = \eta^{\alpha\lambda}F_{\lambda\nu}(A)$ (see formulas (4.25) and (4.26)). For $k = q/mc^2$, Eq. (5.4) is the equation of motion of a particle with charge-to-mass ratio q/m in an electromagnetic field with antisymmetric field strength tensor F_ν^α (see also [61, Eq. (12.3)]). Hence, A is a well-defined four-potential of the electromagnetic field.

The connection between tensor F_ν^α and the more familiar 3D electric field $\mathbf{E} = (E_1, E_2, E_3)$ and 3D magnetic field $\mathbf{B} = (B^1, B^2, B^3)$ can be seen in the matrix forms of the tensor:

$$F_{\mu\nu} = \begin{pmatrix} 0 & E_1 & E_2 & E_3 \\ -E_1 & 0 & -cB^3 & cB^2 \\ -E_2 & cB^3 & 0 & -cB^1 \\ -E_3 & -cB^2 & cB^1 & 0 \end{pmatrix} \quad \text{and} \quad F_\nu^\mu = \begin{pmatrix} 0 & E_1 & E_2 & E_3 \\ E_1 & 0 & cB^3 & -cB^2 \\ E_2 & -cB^3 & 0 & cB^1 \\ E_3 & cB^2 & -cB^1 & 0 \end{pmatrix}. \tag{5.5}$$

Hence, for $j \in \{1, 2, 3\}$, we have

$$E_j = F_{0j} = -F_{j0}, \quad cB^j = -\frac{1}{2}\varepsilon^{jkl}F_{kl}, \tag{5.6}$$

where ε is the *rank* 3 *Levi-Civita pseudoscalar* defined by

$$\varepsilon_{jkl} = \varepsilon^{jkl} = \begin{Bmatrix} 1, & \text{if } (jkl) \text{ is an even permutation of } 123 \\ -1, & \text{if } (jkl) \text{ is an odd permutation of } 123 \\ 0, & \text{otherwise} \end{Bmatrix}. \tag{5.7}$$

We point out that the usual cross product of two vectors $\mathbf{x}, \mathbf{y} \in \mathbb{R}^3$ can be defined using the Levi-Civita pseudoscalar:

$$(\mathbf{x} \times \mathbf{y})_j = \varepsilon_{jkl} x^k y^l = \frac{1}{2}\varepsilon^{jkl}(x_k y_l - x_l y_k). \tag{5.8}$$

We check now that the tensor $F_{\mu\nu}$ produces the correct acceleration of a charge q of mass m due to the Lorentz force. This 3D acceleration is ([61, p. 260, Eq. (6.113)])

$$\mathbf{a} = \frac{q}{m}(\mathbf{E} + \mathbf{v} \times \mathbf{B}). \tag{5.9}$$

Now, since $\dot{x} = \gamma(1, \mathbf{v}/c)$, with γ defined by (3.19), Eq. (5.4) with $j = 1$ implies that

$$\ddot{x}^1 = k\gamma\left(E_1 + B^3 v^2 - B^2 v^3\right) = k\gamma(\mathbf{E} + \mathbf{v} \times \mathbf{B})^1.$$

The cases $j = 2, 3$ are similar, and we obtain $\ddot{x}^j = k\gamma(\mathbf{E} + \mathbf{v} \times \mathbf{B})^j$. For motion with low velocities $v \ll c$, we have $\tau \approx ct$ and $\gamma = 1$ for order less than $(v/c)^2$. Thus, $\ddot{\mathbf{x}} \approx k(\mathbf{E} + \mathbf{v} \times \mathbf{B})$. For

$$k = \frac{q}{mc^2}, \tag{5.10}$$

we have $\ddot{\mathbf{x}} \approx \frac{1}{c^2}\mathbf{a}$, and $F_{\mu\nu}$ produces the correct acceleration due to the Lorentz force. Therefore, Eq. (5.4), with k defined by (5.10), is indeed the relativistic equation of motion of a charge in an electromagnetic field defined by the tensor F^{α}_{ν}.

If we perform a space inversion, the components of $\mathbf{E}$ will change sign, while the components of $\mathbf{B}$ will not change. These are well-known properties of electric and magnetic fields. A word of warning, however: the three-dimensional arrays $\mathbf{E} = (E_1, E_2, E_3)$ and $\mathbf{B} = (B^1, B^2, B^3)$ are *not* tensors. They are neither vectors nor covectors. They do not transform between inertial frames like vectors or covectors. They are *part* of the tensor $F_{\mu\nu}$. As such, there is no significance to the placement of the indices on the components of $\mathbf{E}$ and $\mathbf{B}$. We have chosen to use lower indices on $\mathbf{E}$ because the electric force on a test charge q is $\mathbf{F} = q\mathbf{E}$, and the force $\mathbf{F}$ is a covector. We use upper indices on $\mathbf{B}$ so that the cross product $\mathbf{v} \times \mathbf{B}$ will have lower indices.

5.2 The Four-Potential of a Single-Source Electric Field

Consider the field generated by a single moving source of charge Q. Let $\psi(\tau)$ denote the worldline of the source. For a given spacetime position x, define the retarded time $\tau(x)$ such that $(x - \psi(\tau(x)))^2 = 0$ and $x^0 - \psi^0(\tau(x)) > 0$. The point $\psi(\tau(x))$

is the intersection of the backward light cone with vertex x and the worldline $\psi(\tau)$. Since the field propagates with the speed of light, the position of the source at the retarded time is the only influence on the object at x (see Fig. 3.14). Let $r(x)$ denote the relative position null four-vector:

$$r(x) = x - \psi(\tau(x)). \tag{5.11}$$

Denote the four-velocity of the source at the retarded time by $w(\tau(x))$.

The Lorentz-invariant description of the geometric influence of the field is defined by the covector-valued function $A_\mu(x)$. As shown in Sect. 3.8, $A_\mu(x)$ has one of the two forms

$$\tilde{A}_\mu(x) = g(r \cdot w) r_\mu(x) \tag{5.12}$$

and

$$A_\mu(x) = h(r \cdot w) w_\mu(\tau(x)), \tag{5.13}$$

where g and h are scalar functions of the Lorentz-invariant scalar $r \cdot w$. We identify the functions g and h using the Newtonian limit, that is, by comparing the equation of motion (5.4) with the classical formula when the source and the test charge are at rest.

Let Q be a charge at rest at the origin. Classically, the 3D acceleration, due to Q, of a charge q of mass m at rest at the point $\mathbf{x} = (x^1, x^2, x^3)$ is, by formula (2.5),

$$a^j = \frac{1}{4\pi\epsilon_0} \frac{q}{m} \frac{Qx^j}{|\mathbf{x}|^3}, \tag{5.14}$$

where $|\mathbf{x}| = \sqrt{(x^1)^2 + (x^2)^2 + (x^3)^2}$ is the length of $\mathbf{x}$ and the distance of $\mathbf{x}$ from the origin. Since our charge q is at rest, by (5.2) $\tau = ct$, $\dot{x}^\nu = (1, 0, 0, 0)$ and $a^j = c^2 \ddot{x}^j$ for $j = 1, 2, 3$. Then (5.4) and (5.10) imply that $A_{0,j} - A_{j,0} = \frac{1}{4\pi\epsilon_0} \frac{Qx_j}{|\mathbf{x}|^3}$. Note that the field is time independent, implying that A_μ is also time independent, so $A_{j,0} = 0$. Thus,

$$A_{0,j} = \frac{1}{4\pi\epsilon_0} \frac{Qx_j}{|\mathbf{x}|^3} = \frac{1}{4\pi\epsilon_0} \frac{Q}{|\mathbf{x}|^2} \frac{x_j}{|\mathbf{x}|}. \tag{5.15}$$

Since the source Q is at rest, we have $w(\tau(x)) = (1, 0, 0, 0)$ and $r(x) = (|\mathbf{x}|, \mathbf{x})$. Thus, $r(x) \cdot w(\tau(x)) = |\mathbf{x}|$.

Now assume that A_μ has the form (5.13). Then $A_\mu = h(|\mathbf{x}|)(1, 0, 0, 0)$. Since $\partial_j |\mathbf{x}| = \frac{x^j}{|\mathbf{x}|} = -\frac{x_j}{|\mathbf{x}|}$, we have $A_{0,j} = \partial_j h(|\mathbf{x}|) = -h'(|\mathbf{x}|)\frac{x_j}{|\mathbf{x}|}$, and it follows from (5.15) that $h'(|\mathbf{x}|) = -\frac{1}{4\pi\epsilon_0} \frac{Q}{|\mathbf{x}|^2}$. Since $h(|\mathbf{x}|)$ must vanish at infinity, we obtain $h(|\mathbf{x}|) = \frac{1}{4\pi\epsilon_0} \frac{Q}{|\mathbf{x}|}$. Thus, $A = \left(\frac{Q}{4\pi\epsilon_0 |\mathbf{x}|}, 0, 0, 0\right)$ for a rest charge Q at the origin. Note that the zero component A_0 is the classical scalar potential $U^{\mathrm{EM}}(\mathbf{x})$ of the electromagnetic field of a rest charge Q (see Eq. (2.2)).

Using Lorentz covariance, the four-potential A_μ for a moving charge Q is

$$A_\mu(x) = \frac{Q}{4\pi\epsilon_0} \frac{w_\mu(\tau(x))}{r(x)\cdot w(\tau(x))}. \tag{5.16}$$

This is the Lorentz-covariant Liénard-Wiechert four-potential for a field generated by a moving charge (see [43, 55, 61]).

Next, assume that the four-potential has the form (5.12) and that Q is at rest at the origin. Then $\tilde{A}_\mu = g(|\mathbf{x}|)(|\mathbf{x}|, -\mathbf{x})$, and from (5.15), it follows that $-\left(g'(|\mathbf{x}|)|\mathbf{x}| + g(|\mathbf{x}|)\right)\frac{x_j}{|\mathbf{x}|} = \frac{1}{4\pi\epsilon_0}\frac{Q}{|\mathbf{x}|^2}\frac{x_j}{|\mathbf{x}|}$ and $g(|\mathbf{x}|) = \frac{1}{4\pi\epsilon_0}\frac{Q}{(|\mathbf{x}|)^2}$. Thus, $\tilde{A} = \frac{1}{4\pi\epsilon_0}\frac{Q}{|\mathbf{x}|^2}$ $(|\mathbf{x}|, -\mathbf{x}) = \frac{1}{4\pi\epsilon_0}\left(\frac{Q}{|\mathbf{x}|}, \frac{Qx_1}{|\mathbf{x}|^2}, \frac{Qx_2}{|\mathbf{x}|^2}, \frac{Qx_3}{|\mathbf{x}|^2}\right)$. Note that $\tilde{A}_0 = A_0$ is also here the classical scalar potential $U^{\text{EM}}(\mathbf{x})$ of the electromagnetic field of a rest charge. The spatial part of $\tilde{A}$ is $\frac{1}{4\pi\epsilon_0}Q\nabla\ln(|\mathbf{x}|)$, and $\tilde{A}_\mu(x) = A_\mu(x) + \frac{1}{4\pi\epsilon_0}Q\ln(|\mathbf{x}|)_{,\mu}$. This implies that $A_\mu(x)$ and $\tilde{A}_\mu(x)$ define the same tensor $F_{\mu\nu}$ for a field generated by a charge at rest.

Using Lorentz covariance, the function $\tilde{A}_\mu$ for a moving charge Q is

$$\tilde{A}_\mu(x) = \frac{Q}{4\pi\epsilon_0} \frac{r_\mu(x)}{(r(x)\cdot w(\tau(x)))^2}. \tag{5.17}$$

If the source moves with constant velocity, then the four-potential $\tilde{A}_\mu$ defines the same field tensor $F_{\mu\nu}$ as $A_\mu(x)$. However, as we will see in the next section, if the source is accelerating, $\tilde{A}_\mu$ does not produce a radiation field. Since we know that, in general, electromagnetic fields radiate, only the Liénard-Wiechert four-potential (5.16) is appropriate for describing the electromagnetic field of a moving source.

In order to explore the geometry of an object's spacetime, consider the influence of the field generated by a charge Q, at rest at the origin, on a charge with charge-to-mass ratio q/m. Using (5.1) and (5.16), the action $L(x, u)$ at a point $x = (|\mathbf{x}|, \mathbf{x})$ for an arbitrary four-vector u, is

$$L(x, u) = \sqrt{\eta_{\mu\nu}u^\mu u^\nu} + \frac{\alpha u^0}{|\mathbf{x}|}, \tag{5.18}$$

where $\alpha = \frac{Qq}{4\pi\epsilon_0 mc^2}$. From the geometric meaning of the action function $L(x, u)$ as the infinitesimal distance in an object's spacetime between x and $x + \epsilon u$, it is apparent that for a given u^0, the scaling of spatial displacements with respect to Minkowski space is the same in all directions. See Fig. 5.1.

In Fig. 5.2, for $qQ > 0$, the field is repulsive, and the EM hyperboloid

$$\sqrt{(u^0)^2 - (u^1)^2 - (u^2)^2} + \frac{\alpha u^0}{|\mathbf{x}|} = 1$$

where $\alpha = Qq/(4\pi\epsilon_0 mc^2)$, lies between the light cone and the four-velocities and approaches the four-velocities as the distance from the source increases. For $qQ < 0$,

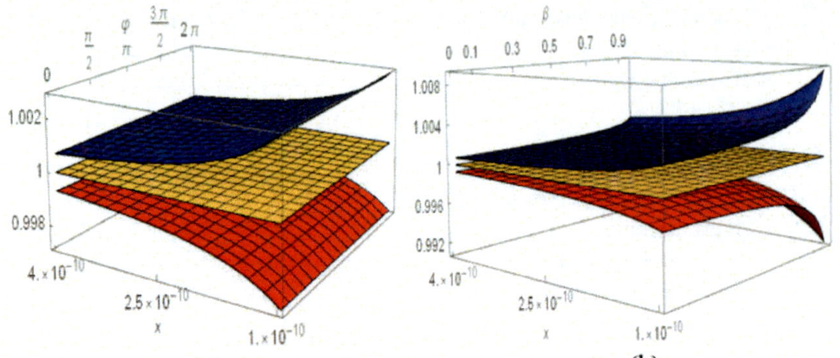

(a) (b)

Fig. 5.1 The action function $L(x, u)$ of Eq. (5.18) for a source at rest, and for $|\alpha| \approx 2.8 \times 10^{-13}$ m, $|\mathbf{x}| \in [10^{-10}, 4 \times 10^{-10}]$ m. **a** The test charges $q = e^+$ (blue), $q = e^-$ (red), and a neutral particle $q = 0$ (yellow) are at rest. In all three cases, the action $L(x, u)$, and hence, the force, does not depend on the spatial direction of u, and the force is stronger nearer the source. For e^+, the negative gradient $-\partial L/\partial x > 0$, and so the force is repulsive. For e^-, the negative gradient $-\partial L/\partial x < 0$, and so the force is attractive. This coincides with (5.14). **b** The three test charges have common four-velocity $\gamma(1, \boldsymbol{\beta})$, where $\gamma = (1 - \beta^2)^{-1/2}$. For varying β. The geometry here depends only on β and not on the spatial direction of u. The test charge is more influenced when its velocity is greater. See Fig. 5.5 for the geometry near a moving source

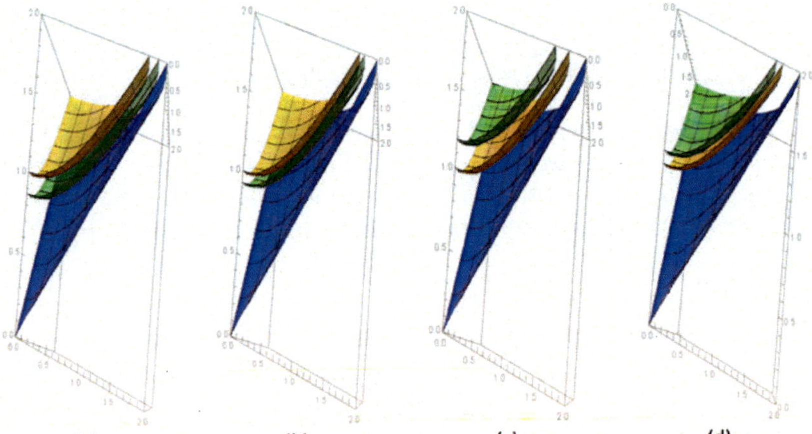

(a) (b) (c) (d)

Fig. 5.2 The influence of an electric field, generated by a single source $Q \approx \pm 1500$ C on Minkowski space. Each plot shows a cutaway of the light cone (in blue), the hyperboloid of four-velocities (in yellow), and the upper half of the hyperboloid $\sqrt{(u^0)^2 - (u^1)^2 - (u^2)^2} + \frac{\alpha u^0}{|\mathbf{x}|} = 1$, where $\alpha = Qq/(4\pi\epsilon_0 mc^2)$ (in green). **a** and **b** are for $\alpha \approx 0.15$ mm and $|\mathbf{x}| = 1$ mm and 2 mm, respectively. **c** and **d** are for $\alpha \approx -0.15$ mm, and $|\mathbf{x}| = 1$ mm and 2 mm, respectively

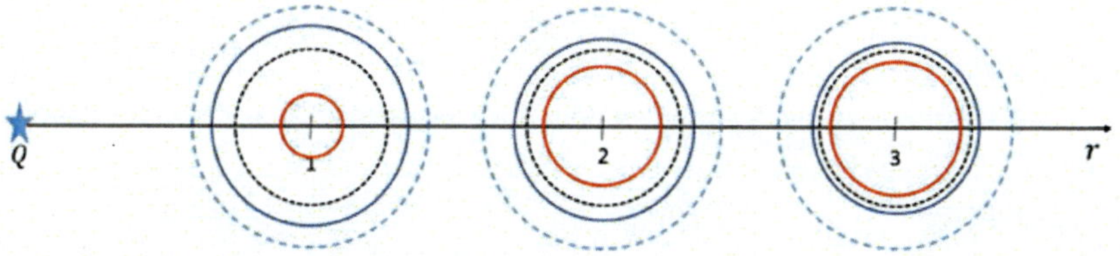

Fig. 5.3 The effect of a source at rest on the geometry of a test charge with charge-to-mass ratio $q/m = 1$, at three different distances $|\mathbf{x}| = 1, 2, 3$ mm. Displayed are the intersection of the plane $u^0 = 1.3$ with the light cone (dashed blue line), the hyperboloid of four-velocities (dashed black line), and two EM hyperboloids $\sqrt{(u^0)^2 - (u^1)^2 - (u^2)^2} + \frac{\alpha u^0}{|\mathbf{x}|} = 1$, where $\alpha \approx 0.15$ mm (blue solid line), and $\alpha \approx -0.15$ mm (red solid line)

the field is attractive, and the EM hyperboloid is inside the four-velocities and also approaches the four-velocities as the distance from the source increases. Figure 5.3 displays horizontal cuts of the light cone, the hyperboloid of four-velocities, and EM hyperboloids. Note that all of the cuts are circles and that the EM circles approach the four-velocities as the distance from the source increases.

5.3 The Electromagnetic Field of a Moving Source

In this section, we compute the electromagnetic field from the Liénard-Wiechert potential $A_\mu(x)$ (5.16). This is the field of a moving charge Q. It has two parts, the usual Coulomb field, which falls off like $1/r^2$, and a far field, which falls off like $1/r$. Next, we make the parallel calculations for the alternative four-potential $\tilde{A}_\mu(x)$ (5.17). The resulting field in this case is the Coulomb field by itself. There is no far field, even if the source is accelerating. This contradicts the fact that, in general, an electromagnetic field radiates and the radiation is part of the far field. Therefore, we conclude that the four-potential $\tilde{A}_\mu(x)$ (5.17) is not appropriate to describe the electromagnetic field of a moving source.

We begin with the Liénard-Wiechert four-potential $A_\mu(x)$ (5.16). Using formulas (3.89) and (3.90), we have

$$A_{\nu,\mu} = \frac{Q}{4\pi\epsilon_0}\left[\frac{w_{\nu,\mu}(r\cdot w) - w_\nu (r\cdot w)_{,\mu}}{(r\cdot w)^2}\right]$$

$$= \frac{Q}{4\pi\epsilon_0}\left[\frac{a_\nu r_\mu}{(r\cdot w)^2} - \frac{w_\nu w_\mu}{(r\cdot w)^2} + \frac{w_\nu r_\mu}{(r\cdot w)^3} - \frac{w_\nu r_\mu (a\cdot r)}{(r\cdot w)^3}\right]. \tag{5.19}$$

To simplify our notations, we define a wedge product of two covectors a and b. The *wedge product* $a \wedge b$ is a rank 2 antisymmetric tensor

$$(a \wedge b)_{\mu\nu} = a_\mu b_\nu - a_\nu b_\mu. \tag{5.20}$$

The action of a wedge product on a four-vector $x \in \tilde{M}$ is a covector defined by

$$((a \wedge b)x)_\mu = (a \wedge b)_{\mu\nu} x^\nu = (a_\mu b_\nu - a_\nu b_\mu) x^\nu. \tag{5.21}$$

This definition implies that

$$(a \wedge b)x = a(b \cdot x) - b(a \cdot x). \tag{5.22}$$

Note that for 3D vectors **a**, **b**, **x**,

$$\mathbf{a}(\mathbf{b} \circ \mathbf{x}) - \mathbf{b}(\mathbf{a} \circ \mathbf{x}) = \mathbf{x} \times (\mathbf{a} \times \mathbf{b}), \tag{5.23}$$

where $\circ$ is the usual Euclidean inner product on $\mathbb{R}^3$.

Using the wedge product and (5.19), the electromagnetic tensor $F_{\mu\nu} = A_{\nu,\mu} - A_{\mu,\nu}$ associated with the Liénard-Wiechert four-potential $A_\mu(x)$ (5.16) is

$$F = \frac{Q}{4\pi\epsilon_0}\left[\frac{r \wedge a}{(r \cdot w)^2} + \frac{r \wedge w}{(r \cdot w)^3} - \frac{(r \wedge w)(a \cdot r)}{(r \cdot w)^3}\right],$$

which can be rewritten as

$$F = \frac{Q}{4\pi\epsilon_0}\left[\frac{r \wedge w}{(r \cdot w)^3} + \frac{(r \wedge a)(r \cdot w) - (r \wedge w)(a \cdot r)}{(r \cdot w)^3}\right].$$

By factoring out an r in the numerator of the second term and using (5.22), we have

$$(r \wedge a)(r \cdot w) - (r \wedge w)(a \cdot r) = r \wedge [a(w \cdot r) - w(a \cdot r)] = r \wedge (a \wedge w)r.$$

Therefore, the field is

$$F = \frac{Q}{4\pi\epsilon_0}\left[\frac{r \wedge w}{(r \cdot w)^3} + \frac{r \wedge (a \wedge w)r}{(r \cdot w)^3}\right]. \tag{5.24}$$

The first term in (5.24) is the Coulomb field

$$F_C = \frac{Q}{4\pi\epsilon_0}\frac{r \wedge w}{(r \cdot w)^3}. \tag{5.25}$$

If the source does not accelerate, the second term vanishes, and the field generated by the source is the Coulomb field. It falls off at large distances like $1/r^2$.

The second term

$$F_r = \frac{Q}{4\pi\epsilon_0}\frac{r \wedge (a \wedge w)r}{(r \cdot w)^3} \tag{5.26}$$

in (5.24) is the called the *far field* because it falls off at large distances like $1/r$. If there is radiation, there must be a change in the energy of the source. Therefore, the source must be accelerating. Hence, radiation is part of the far field.

Since $F_{\mu\nu}(x) = A_{\nu,\mu}(x) - A_{\mu,\nu}(x)$, any four-potential that differs from A_μ by the gradient of some scalar function, called a *gauge*, produces the same electromagnetic field tensor $F_{\mu\nu}$. This so-called *gauge invariance* is used to simplify the solutions of problems.

A common choice of gauge is the *Lorentz gauge*. We say that the four-potential $A_\mu(x)$ satisfies the Lorentz gauge condition if

$$A^{\mu}{}_{,\mu} = \eta^{\mu\nu} A_{\nu,\mu} = 0. \tag{5.27}$$

Since $\eta^{\mu\nu} a_\nu r_\mu = a \cdot r$, $\eta^{\mu\nu} w_\nu r_\mu = w \cdot r$ and $\eta^{\mu\nu} w_\nu w_\mu = 1$, it follows from formula (5.19) that our four-potential satisfies the Lorentz gauge condition.

We now compute the field

$$\tilde{F}_{\mu\nu} = \tilde{A}_{\nu,\mu} - \tilde{A}_{\mu,\nu}$$

for the alternative four-potential $\tilde{A}_\mu(x)$ (5.17). Using formula (3.88) and (3.90), we have

$$\tilde{A}_{\nu,\mu} = \frac{Q}{4\pi\epsilon_0}\left[\frac{r_{\nu,\mu}(r\cdot w)^2 - 2r_\nu(r\cdot w)(r\cdot w)_{,\mu}}{(r\cdot w)^4}\right]$$

$$= \frac{Q}{4\pi\epsilon_0}\left[\frac{\eta_{\mu\nu}}{(r\cdot w)^2} - \frac{w_\nu r_\mu}{(r\cdot w)^3} - \frac{2r_\nu w_\mu}{(r\cdot w)^3} + \frac{2r_\mu r_\nu}{(r\cdot w)^4} - \frac{2r_\mu r_\nu (a\cdot r)}{(r\cdot w)^4}\right].$$

It is straightforward to check that $\tilde{A}$ does not satisfy the Lorentz gauge.

When we compute the field $\tilde{F}_{\mu\nu}$, the $\eta_{\mu\nu}$ terms will cancel, being symmetric in μ and ν. The same is true for the $r_\mu r_\nu$ terms. Thus,

$$\tilde{F}_{\mu\nu} = \frac{Q}{4\pi\epsilon_0}\frac{r\wedge w}{(r\cdot w)^3},$$

which is the same as the Coulomb field (5.25). Note that $\tilde{A}_\mu(x)$ does not produce a far field, even if the source is accelerating. Since we know that, in general, an electromagnetic field produces radiation, only the four-potential $A_\mu(x)$ (formula (5.16)), and not $\tilde{A}_\mu(x)$, is appropriate for describing the electromagnetic field of a moving source. It is an open question whether there exist fields described by $\tilde{A}$.

5.4 The Electric and Magnetic Components of the Field of a Uniformly Moving Source

Let K be an inertial frame. We consider the field of a single charge Q moving uniformly with four-velocity $w = \gamma(1, \mathbf{w}/c)$ in K. Since the source moves uniformly, its acceleration is 0, so there is no radiation field. Using formula (5.25), the field due to Q at a spacetime position x is $F(x) = \frac{Q}{4\pi\epsilon_0}\frac{r(x)\wedge w}{(r(x)\cdot w)^3}$, where $r(x)$, defined by (3.82) is the relative position null vector between x and Q at the retarded time. Here we will decompose $F(x)$ into its more familiar components – the electric field $\mathbf{E}(x)$ and the magnetic field $\mathbf{B}(x)$.

We decompose r and w as

$$r = r^\mu = r^0(1, \mathbf{n}), \quad w = w^\mu = \gamma(1, \boldsymbol{\beta}), \quad \boldsymbol{\beta} = \frac{\mathbf{w}}{c}, \tag{5.28}$$

where $\mathbf{n}$ is a unit 3D vector in the direction of propagation of the field. In this notation, we have

$$r \cdot w = r^0\gamma(1 - \mathbf{n} \circ \boldsymbol{\beta}). \tag{5.29}$$

Note that in the inertial frame comoving with the source, $\boldsymbol{\beta} = 0$, and so in this frame, $r \cdot w = r^0$ is the distance from the point at which we wish to measure the field to the source's retarded position.

Using (5.6) and (5.20), the electric field components are

$$E_j = F_{0j} = \frac{Q(r_0 w_j - r_j w_0)}{4\pi\epsilon_0 (r^0)^3\gamma^3(1 - \mathbf{n}\circ\boldsymbol{\beta})^3} = \frac{Q(\beta_j - n_j)}{4\pi\epsilon_0 (r^0)^2\gamma^2(1 - \mathbf{n}\circ\boldsymbol{\beta})^3}.$$

After raising the index j, the 3D vector $\mathbf{E}$ is

$$\mathbf{E} = \frac{Q}{4\pi\epsilon_0 (r\cdot w)^2}\frac{\mathbf{n} - \boldsymbol{\beta}}{1 - \mathbf{n}\circ\boldsymbol{\beta}} = \frac{Q}{4\pi\epsilon_0 (r^0)^2\gamma^2}\frac{\mathbf{n} - \boldsymbol{\beta}}{(1 - \mathbf{n}\circ\boldsymbol{\beta})^3}. \tag{5.30}$$

We have written $\mathbf{E}$ in two forms, because each one has its advantages in certain situations. Note that the electric field is radial and points outward from the source's *current* position. See Fig. 5.4 for an explanation of why this is so.

The magnetic field components are

$$cB^j = -\frac{1}{2}\varepsilon^{jkl}F_{kl} = -\frac{Q}{8\pi\epsilon_0}\frac{\varepsilon^{jkl}(n_k\beta_l - n_l\beta_k)}{(r^0)^2\gamma^2(1 - \mathbf{n}\circ\boldsymbol{\beta})^3} = \frac{Q}{4\pi\epsilon_0}\frac{(\boldsymbol{\beta}\times\mathbf{n})^j}{(r^0)^2\gamma^2(1 - \mathbf{n}\circ\boldsymbol{\beta})^3}, \tag{5.31}$$

and so, as a 3D vector,

$$c\mathbf{B} = \frac{Q}{4\pi\epsilon_0 (r\cdot w)^2}\frac{\boldsymbol{\beta}\times\mathbf{n}}{1 - \mathbf{n}\circ\boldsymbol{\beta}}. \tag{5.32}$$

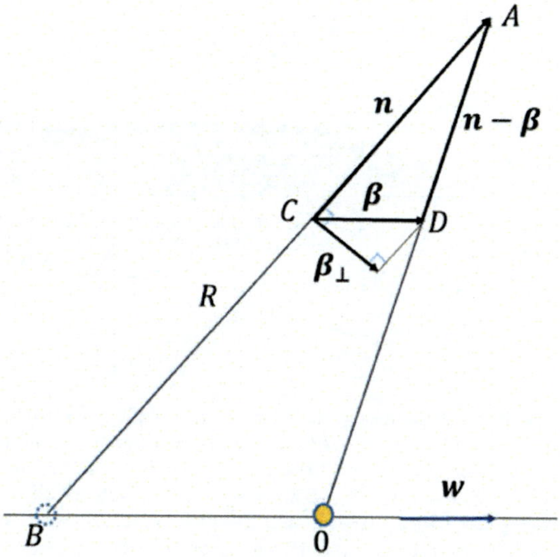

Fig. 5.4 The direction of the electric field **E** at A at time t_0 of a charge moving with constant velocity **w**. The field that reached A at time t_0 was emitted by the charge at position B, at the retarded time $t_0 - \Delta t$ such that $R = |BA| = c\Delta t$. The unit vector **n** is the direction of propagation of the field. Denote by O the position of the source at time t_0. The distance $|BO|$ traveled by the source during the retardation time is $w\Delta t$. From the similarity of triangles $\triangle BAO$ and $\triangle CAD$, it follows that $|CD| = w/c = \beta$. Thus, the direction $\mathbf{n} - \boldsymbol{\beta}$ of the field **E** at A at time t_0 is the direction from the source positioned at O at t_0. The unit-free 3D velocity $\boldsymbol{\beta}$ is decomposed into components $\boldsymbol{\beta}_{\|}$, parallel to **n**, and $\boldsymbol{\beta}_{\perp}$, perpendicular to **n**. The magnetic field at A is proportional to $\mathbf{n} \times \boldsymbol{\beta} = \mathbf{n} \times \boldsymbol{\beta}_{\perp}$ and to the angular momentum of the source with respect to A. See Sect. 5.5

Comparing this formula for $c\mathbf{B}$ with (5.30), we have the following relationships between the electric and the magnetic fields:

$$c\mathbf{B} = \mathbf{n} \times \mathbf{E} = \boldsymbol{\beta} \times \mathbf{E}. \tag{5.33}$$

These relationships agree with [61], Sect. 14.1.

Here, we have calculated the field in the frame K in which the point A is at rest. Alternatively, one could start with the field in the frame K' which is comoving to the source. In K', the field is the Coulomb field, since the source is at rest in K'. Now use the Lorentz transformation from K' to K to transform the field to K. Explicit, if $F'_{\alpha\beta}$ is the field in K', and $L : K' \to K$ is the Lorentz transformation, then the field in K is

$$F_{\mu\nu} = F'_{\alpha\beta} L^{\beta}_{\ \nu} L^{\alpha}_{\ \mu}. \tag{5.34}$$

In order to explore the geometry of the field, we now derive analogs of Eq. (5.18) and Figs. 5.1 and 5.3. Suppose a source charge is moving with four-velocity $w = \gamma(1, \beta, 0, 0, 0)$. Since the field propagates with the speed of light, and the distances

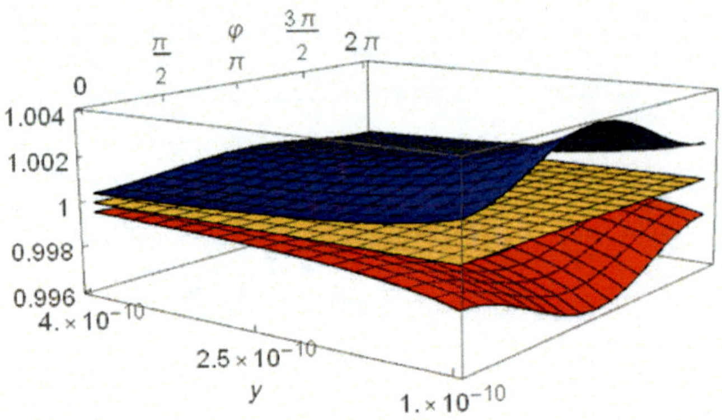

Fig. 5.5 The effect of a source $Q = 100e^+$ moving uniformly at 40% of the speed of light on a test charge $q = e^+$ (blue) and e^- (red). Displayed are the graphs of the action functions $L(y,\varphi) = 1 \pm \alpha(1 - 0.4\cos\varphi)/y$, for $\alpha \approx -2.8 \times 10^{-13}$ m, and $0 \le \varphi < 2\pi$. The distance y from the source ranges from 10^{-10} to 4×10^{-10} m. The yellow surface is the plane $L = 1$ of Minkowski space. Here, unlike Fig. 5.1, $L(y,\varphi)$ depends on φ. The blue surface represents the geometry near a positive test charge e^+. The red surface represents the geometry of a negative test charge e^-. The strength of the electric field is greatest when $\varphi = \pi$, which is the direction antiparallel to the velocity of the source. Here, $|\mathbf{n} - \boldsymbol{\beta}|$ is maximal (see formula (5.30)). Compare to Fig. 5.1

between the source and the test charge are small, we may ignore the retardation. The action function (5.1) at a spatial point $(0, y, 0)$ in the direction φ is then approximately

$$L(y,\varphi) = 1 \pm \alpha \frac{1 - \beta\cos\varphi}{y}. \tag{5.35}$$

See Fig. 5.5.

Using the action function (4.15) and the four-potential (5.16), and setting $w_\mu = \gamma(1, -\beta, 0, 0)$ and $r = r^0(1, \mathbf{n})$, the action $L(x,u)$ at a point $x = (|\mathbf{x}|, \mathbf{x})$ is

$$L(x,u) = \sqrt{(u^0)^2 - (u^1)^2 - (u^2)^2 - (u^3)^2} + \frac{\alpha(u^0 - \beta u^1)}{r^0(1 - n_1\beta)}, \tag{5.36}$$

where $\alpha = \frac{Q}{4\pi\epsilon_0}\frac{q}{mc^2}$. Figure 5.6 displays horizontal sections of EM hyperboloids $L(x,u) = 1$, with $u^0 = 1.3$, $u^3 = 0$, $\beta = 0.6$, and $\alpha \approx 0.15$ mm, at different distances from the source's current position. Compare to Fig. 5.3. When the source is at rest, the EM sections are circles. When the source moves uniformly, the sections are ellipses, shifted in the direction of the motion of the source for negative charges, and shifted in the opposite direction for positive charges. These ellipses approach circles as the distance from the source increases. See Fig. 5.5 for the geometry in the neighborhood of a test charge close to the path of a moving positive source charge, as the source passes nearby.

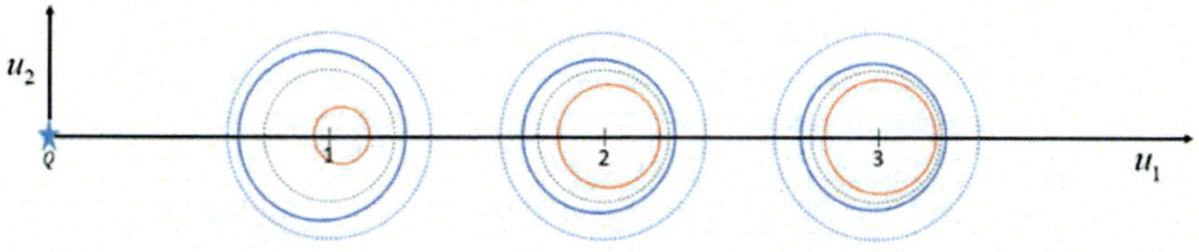

Fig. 5.6 The effect of a source with four-velocity $\approx \gamma(1, 0.6, 0, 0)$ on the geometry of a test charge with charge-to-mass ratio $q/m = 1$, at three different distances $|\mathbf{x}| = 1, 2, 3$ mm from the source's current position. Displayed are the intersection of the plane $u^0 = 1.3$ with the light cone (dashed blue line), the hyperboloid of four-velocities (dashed black line), and the EM hyperboloids $\sqrt{(u^0)^2 - (u^1)^2 - (u^2)^2 - (u^3)^2} + \frac{\alpha(u^0 - \beta u^1)}{r^0(1-n_1\beta)}$, where $\beta = 0.6$, and $\alpha \approx 0.15$ mm (blue solid line), and $\alpha \approx -0.15$ mm (red solid line)

5.5 The Energy-Momentum of an Electromagnetic Field

In this section, we calculate the energy-momentum p^{EM} of a particle moving in the electromagnetic field of a uniformly moving charge Q. We introduce the notion of the *energy-momentum* p^{Q} *of a field*. These energy-momenta will give concrete physical meaning to formulas (5.30) and (5.32) for the electromagnetic field of a moving charge. Moreover, we offer a new interpretation of the magnetic field as a consequence of the angular momentum of the source with respect to the test charge.

From (5.3) and (5.16), the energy-momentum of a test particle of charge q and of mass m in this field is

$$p^{\text{EM}}(x) = kA(x) = \frac{q}{mc^2}\frac{Q}{4\pi\epsilon_0}\frac{w(\tau(x))}{r(x)\cdot w(\tau(x))}, \tag{5.37}$$

where $w = \gamma(1, \boldsymbol{\beta})$ is the four-velocity of the source charge Q. Note that this energy-momentum does not depend on the test particle's velocity.

For a source Q at rest at the origin of an inertial frame, we have

$$p_0^{\text{EM}}(x) = \frac{q}{mc^2}\frac{Q}{4\pi\epsilon_0}\frac{1}{|\mathbf{x}|}\ ,\quad p_j^{\text{EM}} = 0,\ j = 1, 2, 3. \tag{5.38}$$

The physical interpretation is that a source at rest imparts no 3D momentum to the test charge, but it does give the test charge an additional (positive or negative) energy equal to

$$mc^2 p_0^{\text{EM}} = \frac{qQ}{4\pi\epsilon_0}\frac{1}{|\mathbf{x}|} = qU^{\text{EM}}(x), \tag{5.39}$$

where $qU^{\text{EM}}(x)$ is precisely the classical potential energy of our test charge in the field, and $U^{\text{EM}}(x)$ is the *potential energy of the field* (see formula (2.2) and following).

If $qQ > 0$, then $mc^2 p_0^{\text{EM}}(x)$ is the energy gained by a charge q in being brought from infinity (where it was a free particle) to the point $\mathbf{x}$. This energy can become kinetic energy, should the field expel the charge from the field. If $qQ < 0$, then $mc^2 p_0^{\mathrm{EM}}(x)$ is negative, and its value is the amount of energy needed to free the test charge positioned at $\mathbf{x}$ from the field and render it a free particle.

Consider now the case when the source Q is moving with a constant four-velocity $w = \gamma(1, \boldsymbol{\beta})$. We define the *energy-momentum* $p^{\mathrm{Q}}(x)$ *of the field* generated by this source Q by

$$p^{\mathrm{Q}}(x) = \frac{Qw}{4\pi\epsilon_0 c(r \cdot w)} = \frac{Q\gamma(1, \boldsymbol{\beta})}{4\pi\epsilon_0 c(r \cdot w)}. \tag{5.40}$$

This is a new definition of the energy-momentum of a field as a four-vector instead of the usual rank 2 *canonical stress tensor* or its variations (see [61, Sect. 12.10]). Since Q is a Lorentz-invariant scalar, the energy-momentum of the field is a four-vector. This definition of the field energy-momentum is similar to the definition (3.61) of the energy-momentum $\tilde{p}$ of a moving object. As mentioned in the discussion after this definition, the zero component of $\tilde{p}$ is the energy of the object divided by c. Here, too, the zero component is the classical potential energy (2.2) of the field divided by c. Note that the energy-momentum imparted to a test charge is

$$mcp^{\mathrm{EM}}(x) = qp^{\mathrm{Q}}(x). \tag{5.41}$$

The factor γ in the zero component of $p^{\mathrm{Q}}(x)$ expresses the increase in the total energy of the source due to its relativistic kinetic energy. When the speed of the source approaches the speed of light, the energy of the field becomes extremely large. The 3D momentum of the field will also increase. This shows that the energy-momentum gained by a source of an electromagnetic field is transferred to the energy-momentum of the field.

In Fig. 5.4, the carriers of the field that reached the point A at time t_0 were emitted by the source positioned at B at the retarded time. Thus, the direction of the propagation of the field that reached A was $\mathbf{n}$, the direction of the line from B to A. Yet formula (5.30) reveals that the direction of the field $\mathbf{E}$ at A is $\mathbf{n} - \boldsymbol{\beta}$ and not $\mathbf{n}$. The reason for this is that the field imparts a 3D momentum (5.41) to the test charge. Thus, to compute the acceleration felt by the test charge, one must subtract the τ derivative of $\mathbf{p}^{\mathrm{EM}}$, which is proportional to $\boldsymbol{\beta}$, from the direction $\mathbf{n}$ of the force at the retarded time. Figure 5.4 explains why the resultant direction $\mathbf{n} - \boldsymbol{\beta}$ of the acceleration is radial to the source's current position. The magnitude of the acceleration is still proportional to the density of the carriers, which decrease inversely with the square of the distance from the source. This explains formula (5.30) for the electric field of a moving charge.

In order to understand formula (5.32) for the magnetic field, we first recall the classical definitions of orbital angular velocity and orbital angular momentum. Let a particle of mass m have position vector $\mathbf{r}(t)$ with respect to a given point O, not necessarily the origin. For each time t, let $\Pi = \Pi(t)$ be the plane spanned by $\mathbf{r}(t)$ and $\mathbf{v} = d\mathbf{r}/dt$. Use polar coordinates (r, φ) in Π, and let $\omega = \dot{\varphi} = d\varphi/dt$. Then the *orbital angular velocity* of the particle with respect to O is a pseudovector $\boldsymbol{\omega}$ perpendicular to Π of magnitude ω. The orbital angular velocity has units of radians per second and is a pseudovector because it depends on the orientation. To express the orbital angular velocity in terms of the linear velocity $\mathbf{v}$, decompose $\mathbf{v} = \mathbf{v}_{\parallel} + \mathbf{v}_{\perp}$ into components parallel and perpendicular to $\mathbf{r}$. Since $v_{\perp} = r\dot{\varphi}$, we have

$$\boldsymbol{\omega} = \dot{\varphi}\frac{\boldsymbol{\omega}}{\omega} = \frac{v_{\perp}}{r}\frac{\boldsymbol{\omega}}{\omega} = \frac{\mathbf{r} \times \mathbf{v}}{r^2}.$$

The classical *orbital angular momentum* $\mathbf{L}$ of a particle of mass m about O is

$$\mathbf{L} = \mathbf{r} \times \mathbf{p} = \mathbf{r} \times m\mathbf{v} = mr^2\boldsymbol{\omega}. \tag{5.42}$$

Here, $\mathbf{p} = m\mathbf{v}$ is the particle's classical linear momentum.

Now we express the magnetic field (5.32) in terms of p^{Q}. Using (5.29) and (5.40), we have

$$\mathbf{B} = \frac{Q}{4\pi\epsilon_0 c(r \cdot w)^2}\frac{\mathbf{n} \times \boldsymbol{\beta}}{1 - \mathbf{n} \circ \boldsymbol{\beta}} = \frac{\mathbf{r} \times \mathbf{p}^{\mathrm{Q}}}{(r \cdot w)^2}. \tag{5.43}$$

Here, $\mathbf{r}$ is the spatial part of the relative position vector of the source with respect to the test charge, which is the negative of the position vector used in the derivation of the field. This shows that the magnetic field is a multiple of the orbital angular momentum $\mathbf{p}^{\mathrm{Q}}$ of the field with respect to O. The magnitude of the magnetic field is proportional to this angular momentum and to the density of the carriers, which decrease with the square of the distance from the source. This explains formula (5.32).

Until now, we have considered the effect of uniform motion of the source on the field generated by this source. Now what will happen if the source is no longer in uniform motion but *accelerating*? If the velocity of the source is increasing, which happens only when we transfer some additional energy to the source, some of this energy will automatically be transferred to the field. This happens, for example, when we accelerate electrons in the ring of a synchrotron. On the other hand, if we cause a relativistic electron to slow down, the energy and the momentum of the field defined by the source decreases. However, since energy and momentum are conserved, this implies that a part of the energy of the field is no longer connected to the source. This additional energy and momentum is known as *radiant energy* and generates the radiation field, which we discuss in the next section.

5.6 The Radiation Field

In this section, we consider the radiation field (5.26)

$$F_r = \frac{Q}{4\pi\epsilon_0}\frac{r \wedge (a \wedge w)r}{(r \cdot w)^3} \tag{5.44}$$

of a charge Q with four-acceleration $a(\tau)$.

The four-velocity $w(\tau) = \gamma(1, \boldsymbol{\beta}(\tau))$ satisfies $w(\tau) \cdot w(\tau) = 1$, and differentiation by τ yields $a(\tau) \cdot w(\tau) = 0$. By decomposing the four-acceleration as $a = (a^0, \mathbf{a})$, we have $\gamma(a^0 - \mathbf{a} \circ \boldsymbol{\beta}) = 0$ and $a^0 = \mathbf{a} \circ \boldsymbol{\beta}$. Using the decomposition $r = r^0(1, \mathbf{n})$, we have $a \cdot r = r^0(\mathbf{a} \circ \boldsymbol{\beta} - \mathbf{a} \circ \mathbf{n}) = r^0 \mathbf{a} \circ (\boldsymbol{\beta} - \mathbf{n})$. From the definition (5.22) of the wedge product,

$$(a \wedge w)r = a(w \cdot r) - w(a \cdot r) = r^0\gamma[(1 - \mathbf{n} \circ \boldsymbol{\beta})a - \mathbf{a} \circ (\boldsymbol{\beta} - \mathbf{n})(1, \boldsymbol{\beta})]$$

$$= r^0\gamma[\mathbf{n} \circ (\mathbf{n} - \boldsymbol{\beta})a + \mathbf{a} \circ (\mathbf{n} - \boldsymbol{\beta})(1, \boldsymbol{\beta})],$$

which implies that

$$((a \wedge w)r)_j = r^0\gamma[\mathbf{n} \circ (\mathbf{n} - \boldsymbol{\beta})a_j + \mathbf{a} \circ (\mathbf{n} - \boldsymbol{\beta})\beta_j] \quad , \quad j = 1, 2, 3.$$

Denote the Coulomb field by $\mathbf{E}^c$. From (5.30) and (5.44), it follows that the magnetic part of the radiation field is

$$cB^j = -\frac{1}{2}\varepsilon^{jkl}F_{kl}$$

$$= -r^0\frac{1}{2}\varepsilon^{jkl}[(\mathbf{n} \circ \mathbf{E}^c)(n_k a_l - n_l a_k) + (\mathbf{a} \circ \mathbf{E}^c)(n_k \beta_l - n_l \beta_k]$$

$$= -r^0[(\mathbf{n} \circ \mathbf{E}^c)(\mathbf{n} \times \mathbf{a})^j + (\mathbf{a} \circ \mathbf{E}^c)(\mathbf{n} \times \boldsymbol{\beta})^j].$$

Thus,

$$c\mathbf{B} = r^0((\mathbf{n} \circ \mathbf{E}^c)\mathbf{a} + (\mathbf{a} \circ \mathbf{E}^c)\boldsymbol{\beta}) \times \mathbf{n}. \tag{5.45}$$

This shows that the magnetic part $\mathbf{B}$ of the radiation field is transverse to the direction of radiation, that is, $\mathbf{B} \circ \mathbf{n} = 0$.

By property (5.22) of the wedge product, we obtain that $((a \wedge w)r) \cdot r = ((w \cdot r)a - (a \cdot r)w) \cdot r = 0$. Since r is a null vector, this implies that $(r \wedge (a \wedge w)r)r = 0$. Thus,

$$F_{\mu\nu}r^\nu = 0 \tag{5.46}$$

for the radiation field. Using the definition (5.6) of the electric component $\mathbf{E}$, the previous equation for $\mu = 0$ is

$$\mathbf{E} \circ \mathbf{n} = 0. \tag{5.47}$$

Thus, the electric field is also transverse to the direction of propagation.

Similarly, using definition (5.6) of the magnetic components cB^j and substituting $\mu = 1, 2, 3$ in (5.46) yields

$$-\mathbf{E} + c\mathbf{B} \times \mathbf{n} = 0, \quad \text{and } \mathbf{E} = c\mathbf{B} \times \mathbf{n}. \tag{5.48}$$

Using the cross product identity (5.23) and (5.45) yields

$$\mathbf{E} = c\mathbf{B} \times \mathbf{n} = (r^0((\mathbf{n} \circ \mathbf{E}^c)\mathbf{a} + (\mathbf{a} \circ \mathbf{E}^c)\boldsymbol{\beta}) \times \mathbf{n}) \times \mathbf{n}$$

$$= r^0((\mathbf{n} \circ \mathbf{E}^c)(\mathbf{a} \circ \mathbf{n})\mathbf{n} - (\mathbf{n} \circ \mathbf{E}^c)\mathbf{a} + (\mathbf{a} \circ \mathbf{E}^c)(\boldsymbol{\beta} \circ \mathbf{n})\mathbf{n} - (\mathbf{a} \circ \mathbf{E}^c)\boldsymbol{\beta}).$$

Let $\mathbf{a}^\perp$ and $\boldsymbol{\beta}^\perp$, respectively, denote the components of $\mathbf{a}$ and $\boldsymbol{\beta}$ which are orthogonal to $\mathbf{n}$. Then

$$\mathbf{E} = -r^0((\mathbf{n} \circ \mathbf{E}^c)\mathbf{a}^\perp + (\mathbf{a} \circ \mathbf{E}^c)\boldsymbol{\beta}^\perp). \tag{5.49}$$

Finally, (5.48), (5.23) and $\mathbf{B} \circ \mathbf{n} = 0$ lead to

$$\mathbf{n} \times \mathbf{E} = \mathbf{n} \times (c\mathbf{B} \times \mathbf{n}) = c\mathbf{B}. \tag{5.50}$$

As explained above, the radiation field occurs only when the source is accelerating. In this case, $\mathbf{E}$ and $\mathbf{B}$ are perpendicular to the direction of propagation, and $|\mathbf{E}| = c|\mathbf{B}|$. If the source is oscillating with a given frequency, then the fields $\mathbf{E}$ and $\mathbf{B}$ will oscillate with the same frequency, producing plane waves. These waves can be written in the form

$$\mathbf{E}(x) = \mathbf{E}_0 \cos(\omega t - k\mathbf{n} \cdot \mathbf{x}), \tag{5.51}$$

where $\mathbf{E}_0 \circ \mathbf{n} = 0$. The magnetic field $\mathbf{B}$ is defined by (5.50).

5.7 The Four-Potential of a General Electromagnetic Field

The source of an electromagnetic field is a flow of moving charges. We represent this flow in an inertial frame K by a *four-current density*. This is a four-vector

$$J^\mu(x) = \left(\sigma, j^1(x)/c, j^2(x)/c, j^3(x)/c\right), \tag{5.52}$$

where the zero component σ is the *volume charge density* in K, in SI units of coulombs per meter cubed (C/m^3). The spatial vector $\mathbf{j}$ is the current density in K, in SI units of coulombs per second per meter squared ($C/s \cdot m^2$). For example, $j^1(x)$ is the amount of charge per unit time that flows through a unit area perpendicular to the x^1 direction.

To describe the four-current density locally at an arbitrary spacetime point x in K, we assume that in an infinitesimal 3D volume $dv = dx^1dx^2dx^3$ around this point, the charges are moving with the four-velocity $w(x)$. Note that here, the four-velocity $w(x)$ is a function of the spacetime position x and not a function of time alone. We also assume that the total charge in this volume is $Q = \sigma_0 dv$, where σ_0 is called the *rest charge density*. This is the charge density in an inertial frame in which the charges are at rest. This implies that the four-current density (5.52) can be written as

$$J^\mu(x) = \sigma_0 w^\mu(x) \tag{5.53}$$

and that for these charges, $Qw(x) = Jdv$. Therefore, the infinitesimal four-potential (5.16) of the four-current density J^μ is

$$dA_\mu(x) = \frac{J_\mu(x - r(x))dv}{r(x) \cdot w(x - r(x))}. \tag{5.54}$$

We assume that the four-potential of a field of several sources is the sum of their four-potentials. Thus, to define the four-potential of an arbitrary electromagnetic field, we have to integrate the above infinitesimal four-potential $dA_\mu(x)$ over the backward light cone. In Cartesian coordinates, this integration can be challenging, so we will use bipolar coordinates instead. This will greatly simplify the calculations.

Bipolar (BP) coordinates $(\rho^0, \rho^1, \varphi, \theta)$ are defined, for any spacetime point x, by

$$x^0 = \rho^0 \cosh\theta,\quad x^1 = \rho^1 \cos\varphi,\quad x^2 = \rho^1 \sin\varphi,\quad x^3 = \rho^0 \sinh\theta\,, \tag{5.55}$$

where

$$-\infty < \rho^0 < \infty\,,\ 0 \le \rho^1 < \infty\,,\ -\infty < \theta < \infty\,,\ 0 \le \varphi < 2\pi.$$

The angle φ in these coordinates is the usual polar angle φ in the plane $\Pi_1 = \mathrm{Span}(x^1, x^2)$, equipped with the Euclidean metric. In the complementary plane $\Pi_0 = \mathrm{Span}(x^0, x^3)$, the metric is hyperbolic, and so the associated angle θ is hyperbolic.

In *BP* coordinates, a null vector r belongs to one of the hyperplanes $\rho^0 = \pm\rho^1 = \rho$. In particular, all points on the *backward* light cone with vertex at x are of the form $x - r(x) = x - \rho(\cosh\theta, \cos\varphi, \sin\varphi, \sinh\theta)$, where $\rho \ge 0$. On this light cone, the displacements in the spatial coordinates are $dr^1 = d\rho\cos\varphi - \rho\sin\varphi d\varphi$, $dr^2 = d\rho\sin\varphi + \rho\cos\varphi d\varphi$, and $dr^3 = d\rho\sinh\theta + \rho\cosh\theta d\theta$. The 3D volume is the exterior product of these displacements, and we have $dv = \rho^2\cosh\theta d\rho d\theta d\varphi$.

Thus, the four-potential of a field generated by a four-current density $J^\mu(x) = \sigma_0 w^\mu(x)$ is given by integrating (5.54), and

$$A_\mu(x) = \int_0^{2\pi}\int_{-\infty}^{\infty}\int_0^{\infty} \frac{J_\mu(x - r(x))}{r(x) \cdot w(x - r(x))}\rho^2\cosh\theta d\rho d\theta d\varphi, \tag{5.56}$$

with $r(x) = \rho(\cosh\theta, \cos\varphi, \sin\varphi, \sinh\theta)$. This demonstrates that one may calculate the field tensor $F_{\mu\nu}(x) = A_{\nu,\mu}(x) - A_{\mu,\nu}(x)$ from the four-potential of a field generated by a four-current density, without the need for Maxwell's equations.

5.8 The Field of a Current in a Long Wire

Fix an inertial frame K and a spacetime point $z \in K$ inside a conducting wire. The *current* $I = I(z)$ at z is the charge per unit time passing the point. By convention, we consider the positive charges to be at rest in K, while the negative charges move through the wire – in the direction opposite to the current. Current has SI units of coulombs per second. Now imagine a small surface of area A, centered at z, orthogonal to the direction of the current at z. Let I_A denote the current flowing through the surface A. Then the *electric current density* $j(z)$ at z is defined to be

$$j(z) = \lim_{A \to 0} \frac{I_A}{A}.$$

The *current density* $\mathbf{j}(z) = (j^1, j^2, j^3)$ is a 3D vector whose magnitude is $j(z)$ and whose direction is the direction of the current (the positive charges) at z. Thus, $\mathbf{j} = \sigma \mathbf{v}$, where σ is the charge density at z and $\mathbf{v}$ is the velocity of the current. The current density has SI units of coulombs per second per meter squared.

Consider now the field generated by a current moving with constant velocity $\mathbf{w} = (0, 0, w_z)$ in a long wire positioned along the z axis, as in Fig. 5.7. The four-

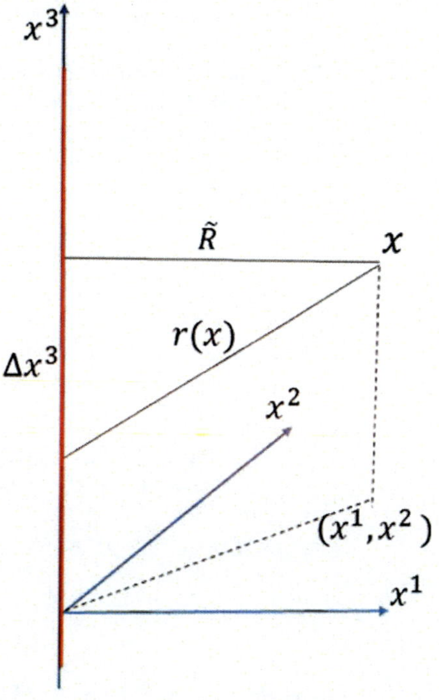

Fig. 5.7 The field of a current in a long wire. The negative charges move down the page, along the x^3 axis. The direction of the current is up the page, along the x^3 axis, so $\mathbf{I} = (0, 0, I)$. $\tilde{R}$ is the distance from x to the wire. $r(x)$ is the spatial part of the relative position null vector

velocity of the current is $w = \gamma(1, 0, 0, w_z/c) = \gamma(1, 0, 0, \beta)$, with $\gamma = \frac{1}{\sqrt{1-\beta^2}}$. For a given point $x = x^\mu$, the relative position vector is $r(x) = (\Delta x^0, x^1, x^2, \Delta x^3)$ and satisfies $(\Delta x^0)^2 = (x^1)^2 + (x^2)^2 + (\Delta x^3)^2$. Let $\tilde{R} = \sqrt{(x^1)^2 + (x^2)^2}$ denote the distance from x to the wire. Fix θ such that $R = \Delta x^0 = \tilde{R}\cosh\theta$ and $\Delta x^3 = \tilde{R}\sinh\theta$. This implies that

$$\mathbf{n} = \frac{1}{\tilde{R}\cosh\theta}(x^1, x^2, \tilde{R}\sinh\theta), \quad \mathbf{n} \circ \boldsymbol{\beta} = \beta\tanh\theta.$$

Denote by σ_0 the rest charge density of the sources of the field, which is assumed to be constant. This means that at a given time, the amount of charge in an interval $[z, z + dz]$ is $Q = \sigma_0 dz$. Using $dz = \tilde{R}\cosh\theta d\theta$, formula (5.30) implies that the $j = 1, 2$ components of the infinitesimal electric field generated by the four-current in $[z, z + dz]$ is

$$dE_j = \frac{\sigma_0(1-\beta^2)n_j}{4\pi\epsilon_0\tilde{R}^2\cosh^2\theta(1-\beta\tanh\theta)^3}dz = \frac{\sigma_0(1-\beta^2)x_j}{4\pi\epsilon_0\tilde{R}^2\cosh^2\theta(1-\beta\tanh\theta)^3}d\theta, \tag{5.57}$$

while

$$dE_3 = \frac{\sigma_0(1-\beta^2)(n_3-\beta)}{4\pi\epsilon_0\tilde{R}^2\cosh^2\theta(1-\beta\tanh\theta)^3}dz = \frac{\sigma_0(1-\beta^2)(\tanh\theta-\beta)}{4\pi\epsilon_0\tilde{R}\cosh\theta(1-\beta\tanh\theta)^3}d\theta. \tag{5.58}$$

Taking the *superposition principle* as an axiom, we obtain that the field generated by any number of sources is the sum of the fields generated by each source. From this principle, it follows that the field generated by the four-current density in the wire is the integral of the fields generated by the four-current density in the infinitesimal intervals dz. Thus, the $j = 1, 2$ components of the field are

$$E_j = \frac{\sigma_0(1-\beta^2)x_j}{4\pi\epsilon_0\tilde{R}^2}\int_{-\infty}^{\infty}\frac{d\theta}{\cosh^2\theta(1-\beta\tanh\theta)^3}.$$

Substitute $u = \tanh\theta$. Then $du = d\theta/\cosh^2\theta$. Since $\tanh\infty = 1 = -\tanh(-\infty)$, the above integral is

$$\int_{-1}^{1}\frac{du}{(1-\beta u)^3} = \frac{1}{2\beta}\left(\frac{1}{(1-\beta u)^2}\right)\Big|_{-1}^{1} = \frac{2}{(1-\beta^2)^2}.$$

Denoting $\tilde{\mathbf{n}} = (x^1, x^2, 0)/\tilde{R}$, we have

$$E_j = \frac{2\sigma_0\gamma^2 x_j}{4\pi\epsilon_0\tilde{R}^2} = \frac{\sigma_0\gamma^2\tilde{n}_j}{2\pi\epsilon_0\tilde{R}}, \quad j = 1, 2. \tag{5.59}$$

Similarly, the third component of the field is

$$E_3 = \frac{\sigma_0(1-\beta^2)}{4\pi\epsilon_0\tilde{R}} \int_{-\infty}^{\infty} \frac{(\tanh\theta - \beta)d\theta}{\cosh\theta(1-\beta\tanh\theta)^3}.$$

The anti-derivative $f(\theta)$ of the integrand can be found using Mathematica. Taking the limits of $f(\theta)$ as θ goes to $\pm\infty$, we find that

$$E_3 = \frac{\sigma_0\gamma\beta}{4\pi\epsilon_0\tilde{R}} \left(\arctan(\gamma(\beta+1)) - \arctan(\gamma(\beta-1))\right), \tag{5.60}$$

which is a finite number.

Consider now the magnetic field $\mathbf{B}$ generated by the current $I = c\sigma_0\gamma\beta$ in our wire. Using (5.33) and $\boldsymbol{\beta} = (0, 0, \beta)$, we have

$$cB_1 = (\boldsymbol{\beta} \times \mathbf{E})_1 = -E_2\beta.$$

Similarly, $cB_2 = E_1\beta$ and $cB_3 = 0$. Now, using (5.59) and $\gamma\beta = I/c\sigma_0$, we have

$$cB_1 = -E_2\beta = -\frac{I\gamma\tilde{n}_2}{2\pi\epsilon_0 c\tilde{R}}, \quad cB_2 = \frac{I\gamma\tilde{n}_1}{2\pi\epsilon_0 c\tilde{R}}.$$

Let $\mathbf{I}$ be the direction of the current. Using the above equations and (2.4), the magnetic field at x is

$$\mathbf{B} = \frac{\mu_0}{2\pi} \frac{\gamma\mathbf{I} \times \mathbf{x}}{(x^1)^2 + (x^2)^2}. \tag{5.61}$$

This agrees with the result given by the Biot-Savart Law ([10], Chap. 3). In fact, the above derivation for a long, straight wire may be modified to yield a *proof* of the Biot-Savart Law.

5.9 Maxwell's Equations

In this section, we derive Maxwell's equations using only the general form (5.6) and (5.5) of the electromagnetic tensor $F_{\mu\nu}$ and Stokes' Theorem.

To obtain the homogeneous Maxwell's equations, first note that the tensor $F_{\mu\nu} = A_{\nu,\mu} - A_{\mu,\nu}$ satisfies the identity

$$F_{\lambda\mu,\nu} + F_{\mu\nu,\lambda} + F_{\nu\lambda,\mu} = 0. \tag{5.62}$$

The proof of this identity is straightforward and uses the symmetry of mixed partial differentiation. For example $A_{\nu,\mu\lambda} = A_{\nu,\lambda\mu}$.

Next, for each $\kappa \in I = \{0, 1, 2, 3\}$, define the index set

$$I_\kappa = I - \{\kappa\}.$$

For $\kappa \in I$, let $I_\kappa = \{\lambda, \mu, \nu\}$. By substituting I_κ in (5.62) for $\kappa \in I$, we obtain four equations. These are precisely the homogeneous Maxwell's equations. To obtain the standard form of these equations, we use the connection (5.6) and (5.5) of the tensor $F_{\mu\nu}$ with the electric and magnetic components **E**, **B** of the field.

For $\kappa = 0$, $\{\lambda, \mu, \nu\} = \{1, 2, 3\}$, and Eq. (5.62) is $F_{12,3} + F_{23,1} + F_{31,2} = 0$. Thus, $-cB^3_{,3} - cB^1_{,1} - cB^2_{,2} = 0$ or

$$\nabla \cdot \mathbf{B} = 0, \tag{5.63}$$

which is Gauss's Law for Magnetism.

For $\kappa = 1$, $\{\lambda, \mu, \nu\} = \{0, 2, 3\}$, and Eq. (5.62) is $F_{02,3} + F_{23,0} + F_{30,2} = 0$. Thus,

$$E_{2,3} - cB^1_{,0} - E_{3,2} = -(\nabla \times \mathbf{E})_1 - cB^1_{,0} = 0.$$

Combining this equation with the equations for $\kappa = 2, 3$, we have

$$\nabla \times \mathbf{E} + c\mathbf{B}_{,0} = \nabla \times \mathbf{E} + \frac{\partial \mathbf{B}}{\partial t} = 0, \tag{5.64}$$

which is the Maxwell-Faraday equation. The homogeneous Maxwell's equations are thus a consequence of the Generalized Principle of Inertia and the fact that the geometry of the electromagnetic field is defined by the *linear* four-potential A_μ (formula (5.16)).

The field tensor $F_{\mu\nu}$ is defined by both the sources of the field, expressed in the four-current density, and the reaction of the surrounding medium, which becomes polarized and magnetized by the sources. To understand the effect of the medium on the field, it is useful to consider an additional tensor which depends only on the sources of the field. The connection between this additional tensor and $F_{\mu\nu}$ will describe the effect of the medium on the electromagnetic field.

The second pair of Maxwell's equations connects the field with its sources, the four-current density $J^\mu(x)$ generating the field. Since charge is conserved, we have $J^\mu(x)_{,\mu} = 0$. Now Stokes' Theorem implies that there is an antisymmetric tensor field $\mathcal{D}^{\mu\nu}(x)$ such that

$$\mathcal{D}^{\mu\nu}_{,\nu}(x) = J^\mu(x). \tag{5.65}$$

This is the tensor form of the non-homogeneous Maxwell's equations. These are four different equations, one for each value of $\mu \in I$.

In the notation of (5.52), the $\mu = 0$ Eq. (5.65) is $\mathcal{D}^{0\nu}_{,\nu}(x) = \rho(x)$. Introducing a 3D vector field $\mathbf{D}(x)$ with components $D^j(x) = \mathcal{D}^{0j}(x) = -\mathcal{D}^{j0}(x)$, we have

$$\nabla \cdot \mathbf{D}(x) = \rho(x), \tag{5.66}$$

which is Gauss's Law.

We will see shortly that $\mathbf{D}(x)$ is closely connected to the electric component **E** of the field. Similarly, there is a pseudovector $\mathbf{H}(x)$, defined by $H_j = \frac{1}{2}\varepsilon_{jkl}\mathcal{D}^{kl}$, which is closely connected to the magnetic component **B**. Now, for $\mu = 1$, Eq. (5.65) is

$$\mathcal{D}^{10}_{,0}(x) + \mathcal{D}^{12}_{,2}(x) + \mathcal{D}^{13}_{,3}(x) = j^1(x),$$

or

$$-D^1_{,0}(x) + H_{3,2}(x) - H_{2,3}(x) = -D^1_{,0}(x) + (\nabla \times \mathbf{H})_1(x) = j^1(x).$$

Combined with the equations for $\mu = 2, 3$, we have

$$\nabla \times \mathbf{H}(x) - \frac{\partial \mathbf{D}}{\partial t}(x) = \mathbf{j}(x), \tag{5.67}$$

which is Ampère's Circuital Law.

We close this section with the connection between **E** and **D** and between **B** and **H**. In vacuum, we have

$$\mathbf{D} = \epsilon_0 \mathbf{E} \quad , \quad \mathbf{B} = \mu_0 \mathbf{H}, \tag{5.68}$$

where ϵ_0 is the permittivity of free space, defined in (2.3), and μ_0 is the permeability of free space, defined by

$$\mu_0 = \frac{1}{c^2 \epsilon_0}. \tag{5.69}$$

In many media, these relationships remain linear, and we have

$$\mathbf{D} = \epsilon \mathbf{E} \quad , \quad \mathbf{B} = \mu \mathbf{H}, \tag{5.70}$$

where ϵ and μ are, respectively, the permittivity and the permeability of the medium. The constants ϵ and μ depend on the medium and how it becomes polarized and magnetized by the free sources. We will return to the subject of electromagnetic fields in media in Chap. 7.

5.10 Orbits of Charged Particles in a Static, Single-source Field

We compute here relativistic orbits of charged particles in a static, single-source electromagnetic field. The derivation is analogous to the derivation of classical orbits in Chap. 2. Orbits in a gravitational field will be handled in Chap. 6.

Consider the motion of a test charge, with charge-to-mass ratio q/m, in a field generated by a charge Q at rest at the origin of the spatial axes. In order to obtain bound orbits, we assume that $qQ < 0$. Let $\mathbf{x}(\tau) \in R^3$ denote the position of the moving charged particle, where τ is defined by (5.2). The relative position null four-vector is $r = (|\mathbf{x}|, \mathbf{x})$. Since the source is at rest, its four-velocity is $w = (1, 0, 0, 0)$. Using (5.4), (5.10) and (5.25), we have, for $j = 1, 2, 3$,

$$\ddot{x}^j = \frac{qQ}{4\pi\epsilon_0 mc^2}\frac{x^j \dot{x}^0}{|\mathbf{x}|^3}.$$

Thus, the 3D acceleration is radial. By Lemma 1 of Sect. 2.2, the motion is in the plane Π generated by the initial position vector $\mathbf{x}(0)$ of the test charge and its initial velocity $\mathbf{v}(0)$. Introduce polar coordinates ρ, φ in the plane Π. By the same lemma, angular momentum is conserved. Using polar coordinates ρ, φ in Π, the angular momentum has constant magnitude

$$l = \rho^2\dot{\varphi}, \tag{5.71}$$

and the square $\dot{\mathbf{r}}^2$ of the magnitude of the velocity is

$$\dot{\rho}^2 + \rho^2\dot{\varphi}^2 = \dot{\rho}^2 + \frac{l^2}{\rho^2}. \tag{5.72}$$

Note that the acceleration is *not* radial with respect to t. We leave this as an exercise.

Since w is constant, and $r \cdot w = \rho$, the action function $L(x, u)$, defined by (5.1), with the four-potential A_μ of (5.16), does not depend on t. From this, it follows that the momentum p_0 is conserved. Thus, the total energy E of the particle is conserved, and

$$p_0 = c\dot{t} + kA_0(\rho) = c\dot{t} - \frac{\alpha}{\rho} = \mathcal{E}, \tag{5.73}$$

where $\alpha = -\frac{qQ}{4\pi\epsilon_0 mc^2} > 0$ and $\mathcal{E} = E/mc^2$ is the total dimensionless energy of the particle, which is constant. Define $r_s = 2\alpha$, and we have

$$c\dot{t} = \mathcal{E} + \frac{r_s}{2\rho}. \tag{5.74}$$

As in the classical case, if $r < r_s$, then the escape velocity of the particle from the field is greater than c, and so r_s is the Schwarzschild radius. Note that $c\dot{t} > 0$, and thus, so is $\mathcal{E} + \frac{r_s}{2\rho}$.

Next, divide Eq. (5.2) by $d\tau^2$ to get

$$1 = c^2\dot{t}^2 - \dot{\rho}^2 - \rho^2\dot{\varphi}^2. \tag{5.75}$$

Substitute $c\dot{t}$ from (5.74) and $\rho\dot{\varphi} = l/\rho$ into (5.75) to obtain

$$\left(\mathcal{E} + \frac{r_s}{2\rho}\right)^2 - \dot{\rho}^2 - \frac{l^2}{\rho^2} = 1. \tag{5.76}$$

This can be rewritten as

$$\dot{\rho}^2 + \frac{l^2 - (r_s/2)^2}{\rho^2} - \frac{r_s\mathcal{E}}{\rho} = \mathcal{E}^2 - 1. \tag{5.77}$$

Next, introduce $f(\varphi) = r_s/(2\rho(\varphi))$, the classical potential energy on the trajectory. Using a prime to denote differentiation by φ, we have $\dot{\rho} = -2lf'/r_s$, exactly as in (2.14). Then (5.77) becomes

$$\frac{4l^2}{r_s^2}(f')^2 + \left(\frac{4l^2}{r_s^2} - 1\right) f^2 - 2\mathcal{E} f = \mathcal{E}^2 - 1\,.$$

As in the classical case (2.16), introduce a dimensionless constant

$$\mu = \frac{r_s^2}{2l^2} \tag{5.78}$$

to obtain

$$(f')^2 = -(1-2\mu)\, f^2 + 4\mu\mathcal{E} f + 2\mu(\mathcal{E}^2 - 1). \tag{5.79}$$

The orbit in question will be bounded when this quadratic has two distinct roots, at the periapsis ρ_p and the apoapsis ρ_a. This occurs when the discriminant of the above quadratic is positive, which occurs exactly when $\mu > (1-\mathcal{E}^2)/2$. Hence, for a bounded orbit, we have

$$f_p + f_a = \frac{4\mu\mathcal{E}}{1-2\mu}\,, \quad f_p f_a = \frac{2\mu(1-\mathcal{E}^2)}{1-2\mu}. \tag{5.80}$$

Incidentally, this implies that $0 < \mathcal{E} < 1$. Differentiating Eq. (5.79) by φ and dividing by $2f'$, we obtain

$$f'' + (1-2\mu) f = 2\mu\mathcal{E}. \tag{5.81}$$

This is a second-order differential equation with constant coefficients. If $\mu > 1/2$, then the solutions will involve exponential functions, and the trajectory will be hyperbolic. For an elliptic orbit, we require $\mu < 1/2$. In this case, the solution will involve periodic functions, and the trajectory will be a bound orbit. In fact, we will have $(1-\mathcal{E}^2)/2 < \mu < 1/2$.

For the bounded case, the general solution of (5.81) is

$$f(\varphi) = \frac{2\mu\mathcal{E}}{1-2\mu}(1 + e\cos(\sqrt{1-2\mu}(\varphi - \varphi_0)), \tag{5.82}$$

where φ_0 is the angle of the periapsis and e is a constant. Now $f_p = \frac{2\mu\mathcal{E}(1+e)}{1-2\mu}$ and $f_a = \frac{2\mu\mathcal{E}(1-e)}{1-2\mu}$, and $f_p > f_a > 0$. Thus, $(f_p - f_a)/(f_p + f_a) = e < 1$. Therefore, the orbit

$$\rho(\varphi) = \frac{A}{1 + e\cos(\sqrt{1-2\mu}(\varphi - \varphi_0))} \tag{5.83}$$

is a precessing ellipse, with one focus at the origin, eccentricity e and semi-latus rectum $A = \frac{r_s(1-2\mu)}{4\mu\mathcal{E}}$. Equation (5.83) agrees with [68, Eq. (5-30)]. We will compute the precession below, but first we should compare this trajectory to the corresponding classical trajectory (2.21), which is a non-precessing ellipse. The significant difference is the factor $\sqrt{1-2\mu}$ in front of the polar angle φ. This is what causes the precession.

To compute the precession, we look for an angle φ such that $\sqrt{1-2\mu}(\varphi-\varphi_0) = 2\pi$. Without loss of generality, take $\varphi_0 = 0$. Then $\varphi = 2\pi/\sqrt{1-2\mu}$, so the precession is

$$\frac{2\pi}{\sqrt{1-2\mu}} - 2\pi = 2\pi\left(\frac{1}{\sqrt{1-2\mu}} - 1\right) = 2\pi\,(1+\mu+\cdots-1) \approx 2\pi\mu. \quad (5.84)$$

If gravity could be described by a linear four-potential, the same analysis could be applied for the motion of Mercury around the Sun. Indeed, it is known that Mercury's orbit is a precessing ellipse. However, the observed precession is $3\pi\mu$. Thus, gravitational fields cannot be modeled with a linear four-potential, but, as we show in the next chapter, they are described properly by a quadratic four-potential.

5.11 Circular Orbits

Here, we analyze the particular case of relativistic circular orbits in an electromagnetic field. From (5.83), we have

$$\rho = R = \frac{r_s(1-2\mu)}{4\mu\mathcal{E}}. \quad (5.85)$$

For a circular orbit, (5.76) becomes

$$\left(\mathcal{E} + \frac{r_s}{2R}\right)^2 = 1 + \frac{l^2}{R^2},$$

or

$$\mathcal{E} = \frac{2\sqrt{R^2+l^2} - r_s}{2R}. \quad (5.86)$$

In (5.85), substitute μ from (5.78) and $\mathcal{E}$ from (5.86) to get

$$R = \frac{r_s(1-\frac{r_s^2}{l^2})}{\frac{2r_s^2}{l^2}\frac{2\sqrt{R^2+l^2}-r_s}{2R}} = \frac{R(l^2-r_s^2)}{r_s(2\sqrt{R^2+l^2}-r_s)}.$$

Cancelling R leads to

$$l^2 - r_s^2 = r_s(2\sqrt{R^2+l^2} - r_s),$$

or

$$l^4 - 4r_s^2 l^2 - 4r_s^2 R^2 = 0.$$

Hence,

$$l^2 = 2r_s^2 + 2r_s\sqrt{r_s^2 + R^2}. \tag{5.87}$$

Let $\Omega = d\varphi/dt$ be the angular velocity of the motion of the object with respect to the lab frame time t. The corresponding angular momentum is $R^2\Omega$ and is equal to $l\frac{d\tau}{dt}$. To obtain an expression for $\frac{d\tau}{dt}$, use (5.75) to get

$$\left(\frac{d\tau}{dt}\right)^2 = \frac{c^2R^2}{R^2 + l^2} = \frac{c^2R^2}{R^2 + 2r_s^2 + 2r_s\sqrt{r_s^2 + R^2}}. \tag{5.88}$$

This implies that the square of the lab frame angular momentum is

$$R^4\Omega^2 = l^2\left(\frac{d\tau}{dt}\right)^2 = \frac{c^2R^2\left(2r_s^2 + 2r_s\sqrt{r_s^2 + R^2}\right)}{R^2 + 2r_s^2 + 2r_s\sqrt{r_s^2 + R^2}}.$$

The square of the angular velocity is thus

$$\Omega^2 = \frac{c^2\left(2r_s^2 + 2r_s\sqrt{r_s^2 + R^2}\right)}{R^2\left(R^2 + 2r_s^2 + 2r_s\sqrt{r_s^2 + R^2}\right)}.$$

For R large enough, Ω will be less than the speed of light c. To see this, rewrite the previous equation as

$$\Omega^2 = \frac{2c^2r_s\left(\frac{r_s}{R} + \sqrt{1 + \frac{r_s^2}{R^2}}\right)}{R^3\left(1 + \frac{2r_s^2}{R^2} + \frac{2r_s}{R}\sqrt{1 + \frac{r_s^2}{R^2}}\right)}.$$

Then, as R approaches ∞, Ω^2 approaches $2c^2r_s/R^3$. Since $r_s = -\frac{qQ}{2\pi\epsilon_0 mc^2}$ (see (5.73) and thereafter), we have, for $R \gg r_s$,

$$\Omega^2 \approx -\frac{qQ}{\pi\epsilon_0 mR^3}. \tag{5.89}$$

Thus, the angular velocity Ω falls off like $1/R^{3/2}$, and the linear velocity $R\Omega$ falls off like $1/\sqrt{R}$.

Chapter 6
The Gravitational Field

This chapter is devoted to the dynamics of the gravitational field. The approach of GR, introduced by Einstein in 1915 [26], considered gravity to curve spacetime. Motion is given by geodesics in curved spacetime. The metric describing the curved spacetime from the sources of the field is given by Einstein's field equations. Standard textbooks include [56, 57, 66, 78, 90, 98].

Our treatment of the dynamics in a gravitational field is different. First of all, we work in an inertial frame instead of in curved spacetime. This approach was introduced in [39] and [40]. We also describe motion by geodesics, but our geodesics are with respect to the action function, which can be defined directly from the sources of the field.

We study the relativistic dynamics associated with the action function $L(x, u)$, defined by (4.15), with $A_\mu(x) = 0$. The action function is thus

$$L(x, u) = \sqrt{\eta_{\mu\nu} u^\mu u^\nu - (l_\mu(x) u^\mu)^2}, \tag{6.1}$$

where $\eta_{\mu\nu}$ is the Minkowski metric (3.37), and $l_\mu(x)$ is a Lorentz-covariant dimensionless covector-valued function. Note that replacing $l(x)$ with $-l(x)$ leads to the same action function. Thus, we assume, without loss of generality, that $l_0 > 0$.

The form of the action function (6.1) is based on the metric

$$g_{\mu\nu}(x) = \eta_{\mu\nu} - h_{\mu\nu}(x), \tag{6.2}$$

where $h_{\mu\nu}(x)$ has the form $l_\mu(x) l_\nu(x)$, for some covector $l_\mu(x)$. The justification for this form of $h_{\mu\nu}$ appears in Chap. 4, after (4.14).

We establish that the quadratic four-potential $l_\mu(x)$ describes a gravitational field. We define a notion of the *field energy-momentum* of an object in a gravitational field and show that a gravitational source imparts a nonzero 3D momentum to test objects

Y. Friedman and T. Scarr, *A Novel Approach to Relativistic Dynamics*,
Fundamental Theories of Physics 210,
https://doi.org/10.1007/978-3-031-25214-3_6

even when the source is at rest. This is in stark contrast to the electromagnetic field (see Eqs. (5.3) and (5.16)).

For a spherically symmetric gravitational field, the dynamics ensuing from (6.1) passes all of the classical tests of GR. After showing that our model predicts the correct precession of the perihelion of Mercury and the correct periastron advance of a binary star, we turn to orbits in the strong field regime. We obtain the surprising result that even here, the angular velocity of circular orbits is classical. We also examine hyperbolic-like orbits of massive objects. By passing to a particular limit, we obtain the corresponding orbits for massless particles. This technique leads to the known formulas for gravitational lensing and the Shapiro time delay. We also describe the gravitational waves predicted by our model and show how to extend the theory to gravitational fields with several sources.

6.1 The Gravitational Field of a Stationary, Static, Spherically Symmetric Body and Its Geometry

In this section, we consider the gravitational field outside a stationary, static, spherically symmetric body of mass M, at rest at the origin of an inertial frame. By Newton's well-known Shell Theorem [80], this field is equivalent to the field of a point source of mass M at rest at the origin.

The geometry of the spacetime of an object moving in such a field is described by a covector-valued function $l_\mu(x)$. To define the zero component $l_0(x)$, let us consider the action of the field on a test body of mass m, positioned at rest at $\mathbf{x}$. For this test mass, we have $\dot{x} = (1, 0, 0, 0)$ and so $l_\mu \dot{x}^\mu = l_0$. The unit-free energy-momentum (4.21) with respect to proper time τ is

$$p_\lambda = \frac{\dot{x}_\lambda - l_0(x) l_\lambda}{\sqrt{1 - l_0^2(x)}} \tag{6.3}$$

and

$$p_0 = \frac{1 - l_0^2(x)}{\sqrt{1 - l_0^2(x)}} = \sqrt{1 - l_0^2(x)}.$$

Thus, the energy of our test body in the field is

$$E = mc^2 p_0 = mc^2 \sqrt{1 - l_0^2(x)} = mc^2 - \frac{1}{2} mc^2 l_0^2(x) + \cdots,$$

where the power series expansion is with respect to $l_0^2(x)$, in analogy with the power series expansion of the kinetic energy in (3.63), which is with respect to v/c. The first term is the rest energy of the body, while the second term is the main part of the additional energy, imparted to the body by the gravitational field.

As was the case for the electromagnetic field, it is natural to assume also here that the first term of this additional energy is the Newtonian potential energy of the test body in the field, which is $U = -\frac{GMm}{|\mathbf{x}|}$, with G defined by (2.7). Since at infinity, the field does not influence the spacetime, its potential energy must vanish there. This implies that the potential energy is uniquely defined, not just "up to a gauge." In fact, we will show later that this assumption leads to the correct Newtonian limit. Thus,

$$-\frac{1}{2}mc^2 l_0^2(x) = -\frac{GMm}{|\mathbf{x}|}, \quad \text{and} \quad l_0^2(x) = \frac{2GM}{c^2|\mathbf{x}|} = \frac{r_s}{|\mathbf{x}|}, \tag{6.4}$$

where

$$r_s = \frac{2GM}{c^2} \tag{6.5}$$

is the Schwarzschild radius and $\frac{r_s}{|\mathbf{x}|}$ is the *dimensionless potential energy of the gravitational field* at $\mathbf{x}$. This suggests defining the *relativistic gravitational potential energy of the test mass at* $\mathbf{x}$ to be

$$E^{\mathrm{G}} = mc^2 \left(\sqrt{1 - \frac{r_s}{|\mathbf{x}|}} - 1 \right). \tag{6.6}$$

This formula is the analog of the relativistic kinetic energy (3.65) for a free particle. It is defined for all $|\mathbf{x}| \geq r_s$, which is reasonable, since we are not discussing dynamics inside the Schwarzschild radius. At the Schwarzschild radius itself, $E^{\mathrm{G}} = -mc^2$. Note that E^{G} is always negative, since a test body must gain energy in order to escape the field. Dividing E^{G} by mc^2, we obtain the dimensionless energy of the field

$$\sqrt{1 - \frac{r_s}{|\mathbf{x}|}} - 1. \tag{6.7}$$

From Chap. 3, Sect. 3.8, there are two forms of Lorentz-covariant four-(co)vector-valued functions for $l_\mu(x)$ (see (3.83) and (3.84)). For a source and a test mass at rest, the relative position null four-vector is $r = (|\mathbf{x}|, \mathbf{x})$, and the four-velocity of the source is $w = (1, 0, 0, 0)$. Thus, $r \cdot w = |\mathbf{x}|$. The first candidate for $l(x)$, based on (3.83), is

$$l_\mu(x) = \sqrt{\frac{r_s}{|\mathbf{x}|^3}}(|\mathbf{x}|, \mathbf{x}) = \sqrt{\frac{r_s}{|\mathbf{x}|}}(1, \mathbf{n}), \tag{6.8}$$

where $\mathbf{n} = \mathbf{x}/|\mathbf{x}|$ is a unit covector in the direction of $\mathbf{x}$. A priori, one might have thought to use $-\mathbf{n}$ instead of $\mathbf{n}$ in the spatial part of $l_\mu(x)$. However, as we will see shortly, this leads to non-physical results.

We observe now an important difference between the gravitational field and the electromagnetic field, when each is generated by a single, static source. For the gravitational field, the 3D momentum (that is, the spatial part of the energy-momentum defined by (6.3) and (6.8)) is nonzero, *even when the source is at rest*. For the electromagnetic field, Eqs. (5.3) and (5.16) imply that when the source is at rest, only the zero component, connected with the energy of the field, differs from zero, while the 3D momentum vanishes. Only when the source is moving does the electromagnetic field impart 3D momentum to test charges.

For a single, static gravitational source at rest, the 3D momentum is

$$p_j = \frac{\dot{x}_j - \frac{r_s}{|\mathbf{x}|^2} x_j}{\sqrt{1 - \frac{r_s}{|\mathbf{x}|}}}. \tag{6.9}$$

The first term in the numerator is the test body's momentum p_j^{fr} as a free particle, while the second term is the momentum p_j^{G} imparted by the field. The denominator handles the gravitational time dilation (see formula (4.50) and the surrounding discussion).

The second option for $l(x)$, based on (3.84), is

$$\tilde{l}_\mu(x) = \sqrt{\frac{r_s}{|\mathbf{x}|}}(1, 0, 0, 0). \tag{6.10}$$

In the next section, we will show that gravity based on $\tilde{l}_\mu(x)$ does not predict the correct precession of Mercury and, thus, is not appropriate to describe a gravitational field. Therefore, we abandon $\tilde{l}_\mu(x)$ and explore now the geometry and the relativistic dynamics induced by the *four-potential* $l_\mu(x)$ defined by (6.8).

We check now the Newtonian limit by use the relativistic dynamics *equation of motion* (4.39) for a stationary worldline

$$\ddot{x}^\alpha = -(l \cdot \dot{x}) F^\alpha_\nu(l) \dot{x}^\nu - (l \cdot \dot{x}) F^\lambda_\nu(l) \dot{x}^\nu l_\lambda (l^\perp)^\alpha + l_{\mu,\nu} \dot{x}^\mu \dot{x}^\nu (l^\perp)^\alpha, \tag{6.11}$$

where $F^\alpha_\nu(l) = \eta^{\alpha\lambda}(l_{\nu,\lambda} - l_{\lambda,\nu})$ and $l^\perp = l - (l \cdot \dot{x})\dot{x}$. Note that in the comoving frame, $l^\perp = (0, \mathbf{l})$ is the spatial part of l, as a four-covector.

For the Newtonian limit, we apply this equation to a test object temporarily at rest, whose four-velocity $\dot{x} = (1, 0, 0, 0)$. This implies $l \cdot \dot{x} = l_0$ and $(l^\perp)^j = l^j$. Thus,

$$\ddot{x}^j = -l_0 F^j_0(l) - l_0 F^\lambda_0(l) l_\lambda l^j + l_{0,0} l^j, \quad j = 1, 2, 3. \tag{6.12}$$

Since the action is time independent, $l_{\mu,0} = 0$. From the definition of $l_0(x)$, using that the derivative of $|\mathbf{x}|$ by x^0 is zero and by x^j is $\frac{x_j}{|\mathbf{x}|}$, we have $l_{0,j} = -\frac{1}{2}\sqrt{\frac{r_s}{|\mathbf{x}|^3}} \frac{x_j}{|\mathbf{x}|}$ and $F^j_0(l) = -\frac{1}{2}\sqrt{\frac{r_s}{|\mathbf{x}|^3}} \frac{x^j}{|\mathbf{x}|}$. Thus,

$$\ddot{x}^j = -\frac{r_s x^j}{2|\mathbf{x}|^3}\left(1 - \frac{r_s}{|\mathbf{x}|}\right).$$

This acceleration is with respect to $\tau = ct$ (since the object is temporarily at rest) in our lab frame. But the object's spacetime is influenced by the gravitational field, and its time is t'. Since the time interval dt' in the influenced spacetime is connected to the invariant parameter in the lab frame via (4.50), we have $c^2dt'^2 = d\tau^2 - (l_0 d\tau)^2$, and the 3D acceleration predicted by our model is

$$\mathbf{a}_N = \frac{d^2\mathbf{x}}{dt'^2} = \ddot{x}\frac{d\tau^2}{dt'^2} = \ddot{x}\frac{c^2}{1-l_0^2} = -\frac{GM\mathbf{x}}{|\mathbf{x}|^3}.$$

This acceleration coincides with (2.8), the Newtonian acceleration due to gravity, at all points outside the Schwarzschild radius. Note that GR has the correct Newtonian limit *only far from the Schwarzschild radius.*

To understand the change in the geometry due to gravitation, we observe that the action function (6.1) of a gravitational field is defined only when the expression under the square root is non-negative. Thus, admissible four-velocities of a motion belong to the domain

$$D_x = \{u : \eta_{\mu\nu}u^\mu u^\nu - (l(x)\cdot u)^2 \geq 0\}. \tag{6.13}$$

It is natural to assume that the admissible four-velocities for light (or any massless particle) belong to the boundary

$$\partial D_x = \{u : \eta_{\mu\nu}u^\mu u^\nu - (l(x)\cdot u)^2 = 0\} \tag{6.14}$$

of D_x. This is the collection of four-vectors u such that $L(x, u) = 0$. Note that the second and third terms on the right-hand side of equation of motion (6.11) ensure that the four-velocities remain admissible throughout the motion.

One way to visualize the geometry of spacetime influenced by a gravitational field is to draw the boundary ∂D_x at different spatial positions $\mathbf{x}$ in the field. Let $\phi = \frac{r_s}{|\mathbf{x}|}$ denote the unit-free gravitational potential of the field at this point. From (6.8), we can write $l_\mu(x) = \sqrt{\phi}(1, \mathbf{n})$, where $\mathbf{n} = \mathbf{x}/|\mathbf{x}|$ is the radial direction. Decompose the velocity $\mathbf{v}$ of a light ray into radial and transverse components $\mathbf{v} = v_n\mathbf{n} + v_t\mathbf{e}_t$, where $\mathbf{e}_t$ is a unit vector perpendicular to $\mathbf{n}$. Let $\beta_n = v_n/c$ and $\beta_t = v_t/c$, so that the four-velocity of this light ray is

$$u = \gamma(v)(1, \beta_n\mathbf{n} + \beta_t\mathbf{e}_t) \tag{6.15}$$

and $(l(x)\cdot u)^2 = \gamma^2\phi(1+\beta_n)^2$. By the definition (6.14) of the boundary, we obtain

$$\gamma^2(1 - \beta_n^2 - \beta_t^2 - \phi(1+\beta_n)^2) = 0,$$

or

$$(1+\phi)\beta_n^2 + 2\phi\beta_n + \beta_t^2 = 1 - \phi. \tag{6.16}$$

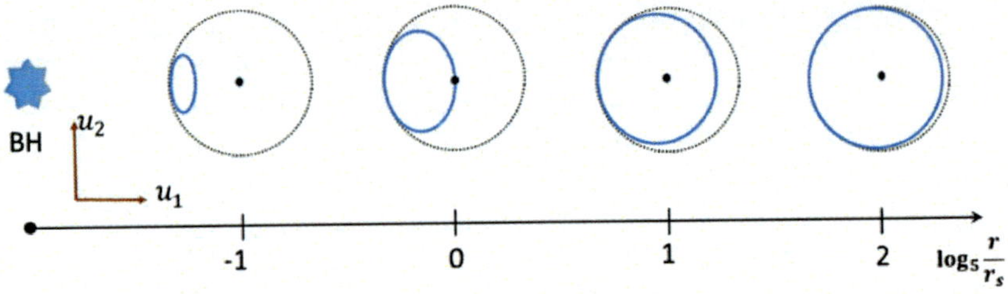

Fig. 6.1 The geometry of spacetime in the vicinity of a black hole. The figure shows the 2D section of the light cone $L(x, u) = 0$ for $u^0 = 1$ in flat spacetime (black, dotted) and influenced spacetime (blue, solid), at several distances $|\mathbf{x}| = r$ from the black hole. The integer k on the scale represents a distance of $5^k r_s$

For example, if the velocity of a massless particle at x is transverse, then

$$\beta_t^2(x) = 1 - \phi(x) = 1 - l_0^2(x) \tag{6.17}$$

Equation (6.16) defines an ellipse in (β_n, β_t) of admissible velocities of light. Direct substitution shows that $\beta_n = -1, \beta_t = 0$ satisfies this equation. This means that in the direction $-\mathbf{n}$ toward the source, the speed of light is c, the speed of light in vacuum. Another solution is $\beta_n = \frac{1-\phi}{1+\phi}, \beta_t = 0$, meaning that in the direction $\mathbf{n}$ away from the source, the speed of light is slowed down by the factor $\frac{1-\phi}{1+\phi}$. This implies that at the Schwarzschild radius, where $\phi = 1$, the speed of light becomes 0, showing that even light cannot escape the region inside the Schwarzschild radius. The center of the ellipse is $C = \left(\frac{-\phi}{1+\phi}, 0\right)$, and the semi-axis associated with β_t is the interval $\left[C, \left(\frac{-\phi}{1+\phi}, \frac{1}{\sqrt{1+\phi}}\right)\right]$.

Now refer to Fig. 6.1. Here we see, at various distances $r = |\mathbf{x}|$ from a black hole, a 2D section of the flat spacetime light cone $L(x, u) = 0$ for $u^0 = 1$, as well as the corresponding section of the light cone in the spacetime influenced by the black hole. Observe that the black hole pulls the light cone toward itself, but leaves it inside the light cone of flat spacetime. At the Schwarzschild radius r_s, the outward component of the light cone is 0, implying that not even light cannot escape this region. Since the influenced light cone is inside the cone of flat spacetime, the speed of a moving object, as observed by an inertial observer, is bounded by the speed of light, *even in presence of a strong gravitational field*.

Another way to describe the geometry is to draw the restriction of the action function to a 2D plane. Consider the action function $L(x, u)$, where $u = \gamma(1, \boldsymbol{\beta})$, where $\boldsymbol{\beta} = (\beta \cos \varphi, \beta \sin \varphi, 0)$. Since $\eta_{\mu\nu} u^\mu u^\nu = 1$, the action function in this case is

$$L(x, u) = \sqrt{1 - \frac{r_s}{|\mathbf{x}|} \gamma^2 (1 + u_r)^2}, \tag{6.18}$$

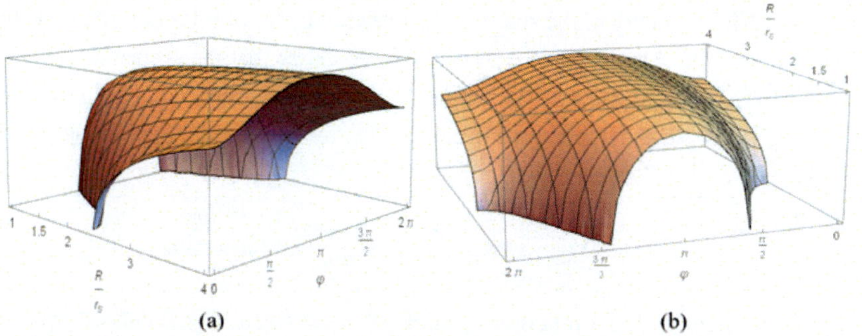

Fig. 6.2 Two perspectives of the action function (6.19), with $\beta = 0.4$, $r_s \le |\mathbf{x}| \le 4r_s$ and $0 \le \varphi < 2\pi$. **a** The shape of the function as $|\mathbf{x}|$ increases is similar to the action function of a *moving* electromagnetic source. This reflects the fact that a gravitational field imparts momentum to objects, even when the source is at rest. In this range, motion is possible in all directions. **b** As one approaches the Schwarzschild radius, the allowable directions of motion become more and more restricted. At the Schwarzschild radius, motion away from the source is impossible

where u_r is the radial component of u. If the spatial part $\mathbf{x}$ of x lies on the positive x axis, then $l_\mu = \sqrt{\frac{r_s}{|\mathbf{x}|}}(1, \mathbf{n}) = \sqrt{\frac{r_s}{|\mathbf{x}|}}(1, 1, 0, 0)$, and the action function is

$$L(x, u) = L(x, \varphi) = \sqrt{1 - \frac{r_s}{|\mathbf{x}|}\gamma^2(1 + \beta \cos \varphi)^2}. \tag{6.19}$$

This function is plotted in Fig. 6.2 for $\beta = 0.4$, $r_s \le |\mathbf{x}| \le 4r_s$ and $0 \le \varphi < 2\pi$.

6.2 Precession of Orbits in a Stationary, Static, Spherically Symmetric Gravitational Field

In the next several sections, we show that our model for the gravitational field of a spherically symmetric body passes all of the tests of GR. In other words, our predictions for the precession of Mercury's perihelion, the periastron advance of a binary star, gravitational lensing, the Shapiro time delay and gravitational waves all match those of GR.

The main tool we will use is the energy-momentum conservation following from the symmetry of the problem and the Euler-Lagrange equations. The energy-momentum of an object moving in the field, in addition to the geometry of its space-time, depends also on the parametrization of its worldline. Moreover, we will not have conservation of angular momentum unless the acceleration is radial with respect to the chosen parameter. It turns out that for the action function (6.1), the acceleration is radial only with respect to the parameter $\tilde{\tau}$, defined by (4.19). Moreover, with this parameter, the momenta have a simpler form.

The geometry of the spacetime influenced by this field is described by the action function (6.1) with $l_\mu(x) = \sqrt{\frac{r_s}{|\mathbf{x}|}}(1, \mathbf{n})$. To check that the acceleration is radial, we first have to calculate the derivatives $l_{\mu,\nu}(x)$ of $l_\mu(x)$. Since $l_\mu(x)$ is time independent, $l_{\mu,0}(x) = 0$, and it is enough to calculate $l_{\mu,j}(x)$, for $j = 1, 2, 3$. Using $|\mathbf{x}|_{,j} = \frac{x^j}{|\mathbf{x}|} = n_j$, we obtain

$$l_{0,j}(x) = \frac{\sqrt{r_s}}{2|\mathbf{x}|^{3/2}} n_j, \quad l_{k,j}(x) = \frac{\sqrt{r_s}}{2|\mathbf{x}|^{3/2}}(2\delta_{kj} - 3n_k n_j) = l_{j,k}(x),$$

for $j, k = 1, 2, 3$. From this, it follows that the tensor $F_{\lambda\nu}(l)$ defined by (4.25) is

$$F_{\lambda\nu}(l) = \{F_{j0} = -F_{0j} = \frac{\sqrt{r_s}}{2|\mathbf{x}|^{3/2}} n_j, \quad 0 \text{ otherwise}\} \tag{6.20}$$

Using formula (4.56) for the acceleration $\frac{d^2x^\alpha}{d\tilde{\tau}^2}$, we observe that the second and third terms connected to the gravitation field are in the direction of l, which is radial, and the first term, which is proportional to $F^\alpha_\nu(l)\dot{x}^\nu$, is also radial by (6.20). Thus, the acceleration is radial, and we have conservation of angular momentum if and only if we use $\tilde{\tau}$ as the evolution parameter.

With respect to $\tilde{\tau}$, the unit-free energy-momentum is

$$p_\lambda(\tilde{\tau}) = \dot{x}_\lambda - (l \cdot \dot{x})l_\lambda, \tag{6.21}$$

where the dot denotes differentiation by $\tilde{\tau}$. Since the acceleration is radial, Lemma 1 of Sect. 2.2 implies that the angular momentum of this motion is conserved, and in the plane of motion,

$$\rho^2\dot{\varphi} = l \tag{6.22}$$

for some constant l. The velocity of a curve $\mathbf{x}(\tau) = (\rho(\tau)\cos\varphi(\tau), \rho(\tau)\sin\varphi(\tau))$ is $\dot{\mathbf{x}} = (\dot{\rho}\cos\varphi - \rho\dot{\varphi}\sin\varphi, \dot{\rho}\sin\varphi + \rho\dot{\varphi}\cos\varphi)$. Introduce an orthogonal spatial basis $\mathbf{e}_\rho = (\cos\varphi, \sin\varphi)$ and $\mathbf{e}_\varphi = (-\sin\varphi, \cos\varphi)$. Using this basis for the spatial components, we have $x = (x^0, \rho, 0)$ and $\dot{x} = (\dot{x}^0, \dot{\rho}, \rho\dot{\varphi})$. Moreover, in this basis, l defined by (6.8) is $l = \sqrt{\frac{r_s}{\rho}}(1, 1, 0)$ and $l_\mu\dot{x}^\mu = \sqrt{\frac{r_s}{\rho}}(\dot{x}^0 + \dot{\rho})$.

The zero component p_0 of (6.21) is proportional to the total energy of the moving object. Since the action function is time independent, p_0 is conserved, implying that

$$p_0 = \left(1 - \frac{r_s}{\rho}\right)\dot{x}^0 - \frac{r_s}{\rho}\dot{\rho} = \mathcal{E}, \tag{6.23}$$

where $\mathcal{E} = E/mc^2$ is the total dimensionless energy of the moving object.

Dividing (4.19) by $d\tilde{\tau}^2$, we find that the square of length of the four-velocity in these coordinates is

$$\left(1-\frac{r_s}{\rho}\right)(\dot{x}^0)^2 - 2\frac{r_s}{\rho}\dot{x}^0\dot{\rho} - \left(1+\frac{r_s}{\rho}\right)\dot{\rho}^2 - \rho^2\dot{\varphi}^2 = 1. \tag{6.24}$$

Substituting in this equation the expressions for $\rho\dot{\varphi}$ and $\dot{x}^0$ from (6.22) and (6.23), respectively, and multiplying by $1-\frac{r_s}{\rho}$, we obtain

$$\left(\mathcal{E}+\frac{r_s}{\rho}\dot{\rho}\right)^2 - 2\frac{r_s}{\rho}\left(\mathcal{E}+\frac{r_s}{\rho}\dot{\rho}\right)\dot{\rho} - \left(1-\frac{r_s^2}{\rho^2}\right)\dot{\rho}^2 - \frac{l^2}{\rho^2}\left(1-\frac{r_s}{\rho}\right) = 1-\frac{r_s}{\rho}. \tag{6.25}$$

Opening the parentheses and simplifying leads to

$$\dot{\rho}^2 + \frac{l^2}{\rho^2}\left(1-\frac{r_s}{\rho}\right) - \frac{r_s}{\rho} = \mathcal{E}^2 - 1\,. \tag{6.26}$$

This equation agrees with [59, Eq. 9.32].

Compare this equation to the corresponding classical Eq. (2.12)

$$\left(\frac{d\rho}{cdt}\right)^2 + \frac{l^2}{\rho^2} - \frac{r_s}{\rho} = \frac{2E}{mc^2}, \tag{6.27}$$

where E is the total energy of the moving object. The sum $\frac{l^2}{\rho^2} - \frac{r_s}{\rho}$ is the sum of the gravitational potential $-\frac{r_s}{\rho}$ and the potential $\frac{l^2}{\rho^2}$ of the centripetal force.

Equation (6.26) is a relativistic extension of (6.27). The radial term $\dot{\rho}^2$ of (6.26) is the classical radial term $(d\rho/dt)^2$ of (6.27) divided by the gravitational time dilation factor $1 - r_s/\rho$. This is as it should be, in order to handle gravitational time dilation. Likewise, the expression $\rho^2\dot{\varphi}^2 = l^2/\rho^2$ in the second term of (6.26) is the classical transverse term $\rho^2(d\varphi/dt)^2$ divided by the gravitational time dilation. As a result, this expression is *multiplied* by the time dilation factor because a *radial* field doesn't affect the transverse direction. Moreover, it is precisely the cubic term ρ^{-3} which is not present in the classical equation and accounts for relativistic effects. We consider

$$V_{\text{eff}} = \frac{l^2}{\rho^2}\left(1-\frac{r_s}{\rho}\right) - \frac{r_s}{\rho} \tag{6.28}$$

to be an "effective potential." We will need this later in the discussion of the stability of orbits.

For some of the problems that we tackle later, it will be convenient to describe the trajectory as $\rho(\varphi)$, which is independent of the parametrization of the worldline. From (6.22), it follows $\frac{d\rho}{d\varphi} = \frac{\dot{\rho}}{\dot{\varphi}} = \frac{\dot{\rho}\rho^2}{l}$. Substituting this into Eq. (6.26) yields

$$\frac{l^2}{\rho^4}\left(\frac{d\rho}{d\varphi}\right)^2 + \frac{l^2}{\rho^2}\left(1 - \frac{r_s}{\rho}\right) - \frac{r_s}{\rho} = \mathcal{E}^2 - 1\,, \tag{6.29}$$

which is an equation for the trajectory $\rho(\varphi)$.

This differential equation becomes simpler if we introduce a function $f(\varphi) = \frac{r_s}{\rho(\varphi)}$, which is the unit-free, classical potential energy on the trajectory. If we denote $f' = \frac{df}{d\varphi}$, then $f' = -\frac{r_s}{\rho^2}\frac{d\rho}{d\varphi}$ and the above equation becomes

$$\frac{l^2}{r_s^2}(f')^2 + \frac{l^2}{r_s^2}(f^2 - f^3) - f = \mathcal{E}^2 - 1\,. \tag{6.30}$$

As in the classical case (2.16) and the relativistic case for electromagnetism (5.78), we introduce a dimensionless constant

$$\mu = \frac{r_s^2}{2l^2} \tag{6.31}$$

and multiply Eq. (6.30) by 2μ to obtain the *planetary orbit equation*

$$(f')^2 = f^3 - f^2 + 2\mu f + 2\mu(\mathcal{E}^2 - 1). \tag{6.32}$$

As shown in Chap. 2, Sect. 2.2, the bounded solution of the classical Newtonian equation (2.17) corresponding to (6.32) is a non-precessing ellipse.

For bounded solutions of (6.32), the maximum and minimum values of f are the roots f_p, f_a of the cubic on the right-hand side of Eq. (6.32). Let $\tilde{\mu} = \frac{1}{2}(f_p + f_a)$. Note that $\tilde{\mu}$ is the average dimensionless potential energy on the trajectory. From the decomposition of a polynomial into a product of first-order terms, it follows that this cubic has an additional root $f_3 = 1 - (f_p + f_a) = 1 - 2\tilde{\mu}$. Hence, we can rewrite (6.32) as

$$(f')^2 = -(f - f_p)(f - f_a)(1 - 2\tilde{\mu} - f). \tag{6.33}$$

To establish the connection between $\tilde{\mu}$ and μ, compare the coefficient 2μ of f in (6.32) with the coefficient in the product of first-order terms of (6.33). This implies that

$$2\mu = f_a f_3 + f_p f_3 + f_a f_p = (f_p + f_a)(1 - (f_p + f_a)) + f_a f_p. \tag{6.34}$$

Using the definition of $\tilde{\mu}$, formula (6.34) leads to the following relationship between μ, f_a and f_p:

$$2\mu = 2\tilde{\mu} - 4\tilde{\mu}^2 + f_a f_p. \tag{6.35}$$

When the trajectory is far from the Schwarzschild radius, we have $f_a, f_p \ll 1$. In this case, the previous equation implies that $\mu \approx \tilde{\mu}$. Until the end of this section, we

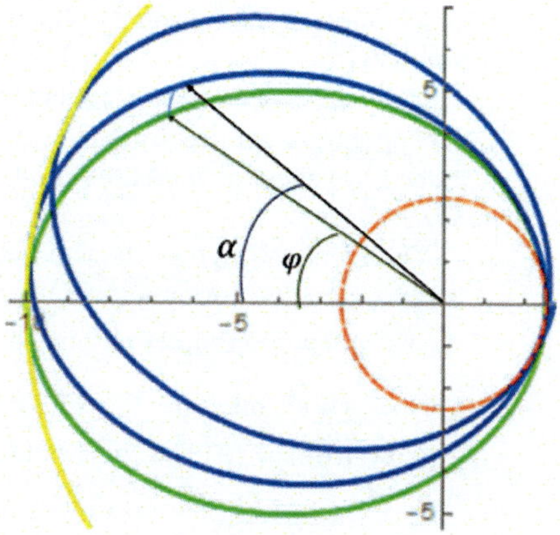

Fig. 6.3 Classical and relativistic orbits of a planet in a stationary, static, spherically symmetric gravitational field. The classical orbit (in green) is a non-precessing ellipse. The relativistic orbit (in blue) is a precessing ellipse

will assume that this is the case. In the next section, we will also consider trajectories near the Schwarzschild radius, that is, in the strong field regime.

We will look for solutions of (6.33) of the form $f(\varphi) = \mu(1 + e\cos\alpha(\varphi))$, where the angle α satisfies $r_{\text{classical}}(\alpha) = r(\varphi)$ (Fig. 6.3). This is no loss of generality, since, on the trajectory, f oscillates between its minimum f_a and its maximum f_p, and the function $\alpha(\varphi)$ is well defined. Substituting this form of $f(\varphi)$ into (6.33), one obtains $d\alpha/d\varphi = \sqrt{1 - 3\mu - \mu e\cos\alpha}$ and the explicit dependence

$$\varphi(\alpha) = \varphi_0 + \int_0^\alpha (1 - 3\mu - \mu e\cos\tilde{\alpha})^{-1/2} d\tilde{\alpha}. \tag{6.36}$$

Thus, the relativistic trajectory of a planet in the gravitational field of a stationary, static, spherically symmetric mass M is

$$\rho(\varphi) = \frac{r_s/\mu}{1 + e\cos\alpha(\varphi)}, \tag{6.37}$$

with $\alpha(\varphi)$ defined by (6.36). The time dependence on the trajectory can be found from (6.22).

Using (6.36) eventually yields the known perihelion precession formula ([78])

$$\varphi(2\pi) - \varphi(0) - 2\pi \approx 3\pi\mu\,\frac{\text{rad}}{\text{rev}}. \tag{6.38}$$

The perihelion of Mercury is $r_p = 4.60012 \times 10^{10}$ m, and its aphelion is $r_a = 6.98168 \times 10^{10}$ m. The Schwarzschild radius of the Sun is $r_s = 2953.25$ m. From $\mu = (f_p + f_a)/2$, we have $\mu = 5.32497 \times 10^{-8}$. Thus, our model predicts a precession of

$$3\pi\mu = 5.01866 \times 10^{-7}\,\frac{\text{rad}}{\text{rev}},$$

which is *exactly* the currently observed precession. Therefore, our model passes this test of GR.

We end this section by showing that the four-potential $\tilde{l}_\mu(x)$, defined by (6.10), is not appropriate for describing the gravitational field of a spherically symmetric body. Indeed, we show now that the orbit precession predicted by $\tilde{l}_\mu(x)$ is $4\pi\mu$ instead of $3\pi\mu$.

The identical arguments as above for l_μ show that also for $\tilde{l}_\mu$, the acceleration is radial and that formula (6.22) holds. As above, the motion is in a plane, and we can use polar coordinates in this plane. However, we now have $\tilde{l} = \sqrt{\frac{r_s}{\rho}}(1, 0, 0)$ and $\tilde{l}_\mu \dot{x}^\mu = \sqrt{\frac{r_s}{\rho}}\dot{x}^0$, implying that (6.23) becomes $p_0 = \left(1 - \frac{r_s}{\rho}\right)\dot{x}^0 = \mathcal{E}$.

Now, from (4.19), the square of length of the four-velocity in these coordinates is

$$\left(1 - \frac{r_s}{\rho}\right)(\dot{x}^0)^2 - \dot{\rho}^2 - \rho^2\dot{\varphi}^2 = 1. \tag{6.39}$$

Substituting in this equation the expressions for $\dot{x}^0$ and $\rho\dot{\varphi}$, one obtains

$$\dot{\rho}^2 + \frac{l^2}{\rho^2} = \frac{\mathcal{E}^2}{1 - \frac{r_s}{\rho}} - 1\,. \tag{6.40}$$

As above, to describe the trajectory, introduce a function $f(\varphi) = \frac{r_s}{\rho(\varphi)}$, which is the unit-free, classical potential energy on the trajectory. Then $\dot{\rho} = -\frac{l}{r_s}f'$ and Eq. (6.40) becomes

$$\frac{l^2}{r_s^2}(f')^2 + \frac{l^2}{r_s^2}f^2 = \frac{\mathcal{E}^2}{1 - f} - 1\,.$$

Multiplying this equation by $\mu = \frac{r_s^2}{2l^2}$, we obtain the *planetary orbit equation*

$$(f')^2 = \frac{f^3 - f^2 + 2\mu f + 2\mu(\mathcal{E}^2 - 1)}{1 - f}, \tag{6.41}$$

whose numerator is the same cubic as in the corresponding Eq. (6.32) for $l_\mu(x)$.

Now factor the right-hand side of (6.41) as

$$(f')^2 = -\frac{(f - f_p)(f - f_a)(1 - 2\mu - f)}{1 - f}$$

and look for solutions of this equation of the form $f(\varphi) = \mu(1 + e\cos\alpha(\varphi))$. Substituting this into the equation, one obtains

$$d\alpha/d\varphi = \sqrt{\frac{1 - 3\mu - \mu e\cos\alpha}{1 - \mu - \mu e\cos\alpha}}$$

and the explicit dependence

$$\varphi(\alpha) = \varphi_0 + \int_0^{\alpha} \left(\frac{1 - 3\mu - \mu e \cos\tilde{\alpha}}{1 - \mu - \mu e \cos\tilde{\alpha}} \right)^{-1/2} d\tilde{\alpha}. \tag{6.42}$$

Expand the integrand as a product of power series, to first order in μ:

$$(1 - 3\mu - \mu e \cos\tilde{\alpha})^{-1/2}(1 - \mu - \mu e \cos\tilde{\alpha})^{1/2}$$

$$= \left(1 + \frac{3}{2}\mu + \frac{1}{2}\mu e \cos\tilde{\alpha} + \cdots\right)\left(1 - \frac{1}{2}\mu - \frac{1}{2}\mu e \cos\tilde{\alpha} - \cdots\right) = 1 + \mu + \cdots .$$

Then, to first order in μ, the precession of the orbit is

$$\varphi(2\pi) - \varphi(0) - 2\pi = \int_0^{2\pi} (1 + \mu) d\tilde{\alpha} - 2\pi = 2\pi\mu \frac{\text{rad}}{\text{rev}},$$

which differs from the observed precession of Mercury orbit. Thus, the four-potential $\tilde{l}_\mu(x)$, defined by (6.10), cannot be used to define the gravitational field of a spherically symmetric body.

6.3 Periastron Advance of Binary Stars

The next test of GR is the periastron advance of a binary star. The *periastron* of a binary system is the closest distance between the two stars as they orbit around each other. Due to the strong gravitational field of the binary, the periastron precesses, and GR accurately predicts the periastron advance of both the Hulse-Taylor binary and the double pulsar PSR J0737-3039A/B. In this section, we compute the periastron advance of a binary star and compare our prediction with that of GR.

Consider now the gravitational field generated by a binary star. Two spherically symmetric objects S_1 and S_2, with respective masses m_1 and m_2, are positioned in an inertial system K, with respective position vectors $\mathbf{r}_1$ and $\mathbf{r}_2$.

The motion of S_1 and S_2 can be reduced, by a standard method (see, for example, [41]) to the motion of a single fictitious planet P in the central field of a Sun of mass $M = m_1 + m_2$, positioned at O.The gravitational force between S_1 and S_2 is along the line joining them (see Fig. 6.4). Denote by $\mathbf{r} = \mathbf{r}_2 - \mathbf{r}_1$ the 3D displacement vector between them and by $\hat{\mathbf{r}}$ the unit vector in the direction $\mathbf{r}$. Let $\mathbf{F} = F\hat{\mathbf{r}}$ be the force acting on S_1, and let $-\mathbf{F}$ be the force acting on S_2. By Newton's Second Law, the accelerations of S_1 and S_2 are

$$\ddot{\mathbf{r}}_1 = \frac{1}{m_1}\mathbf{F}, \quad \ddot{\mathbf{r}}_2 = -\frac{1}{m_2}\mathbf{F}. \tag{6.43}$$

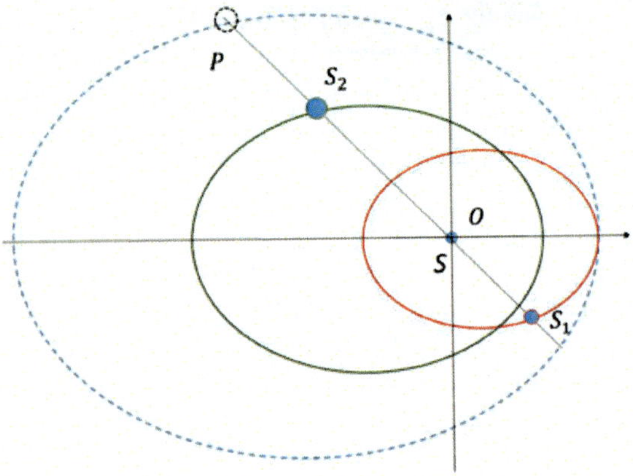

Fig. 6.4 A binary star, consisting of two stars S_1 and S_2, of masses m_1, m_2 respectively, with the center of mass O. Their motion can be reduced to the motion of a single fictitious planet P in the central field of a Sun of mass $M = m_1 + m_2$, positioned at O

From this, the relative acceleration of $\mathbf{r}$ is

$$\ddot{\mathbf{r}} = -\left(\frac{1}{m_1} + \frac{1}{m_2}\right)\mathbf{F} = -\frac{1}{\rho}\mathbf{F}, \tag{6.44}$$

where the reduced mass

$$\rho = \frac{m_1 m_2}{m_1 + m_2}. \tag{6.45}$$

is the mass of a single fictitious object P whose motion satisfies (6.44).

From Eq. (6.43), $m_1\ddot{\mathbf{r}}_1 + m_2\ddot{\mathbf{r}}_2 = 0$, implying that the center of mass defined by

$$\mathbf{R} = \frac{m_1\mathbf{r}_1 + m_2\mathbf{r}_2}{m_1 + m_2} \tag{6.46}$$

moves uniformly. Without loss of generality, we assume that the center of mass is at rest at the origin O of K. Otherwise, we move to an inertial frame in which the center of mass is at rest, do the calculations there, and transform back to K.

Under this assumption, the evolution of the relative separation $\mathbf{r}$ in the two-body problem is reduced to solving the one-body problem

$$\ddot{\mathbf{r}} = -\frac{GM}{r^2}\hat{\mathbf{r}}, \tag{6.47}$$

for the fictitious planet P of mass ρ in the gravitational field of a fictitious sun S of mass $M = m_1 + m_2$, located at the origin O of K. Moreover, since $\mathbf{r} = \mathbf{r}_2 - \mathbf{r}_1$, the apastron separation of the binary (the maximum distance between the two objects) is the same as the aphelion (maximal distance of P from O), and the periastron separation (the minimum distance between the two objects) is the same as the perihelion (minimal distance of P from O). In addition, the eccentricity of the orbit of

P is the same as that of the eccentricity of the orbits of S_1 and S_2. To see this, use $m_1\mathbf{r}_1 + m_2\mathbf{r}_2 = 0$ to get

$$\mathbf{r}_1 = -(m_2/M)\mathbf{r} \quad \text{and} \quad \mathbf{r}_2 = (m_1/M)\mathbf{r}. \tag{6.48}$$

The trajectory $r(\varphi)$ of P is a precessing ellipse, given by (6.37), and its precession is given by (6.38). This is also the periastron advance per revolution of the binary. To compare our results to GR, we compute the commonly used non-Keplerian parameter *periastron advance per unit of time* $\dot{\omega}$, also called the time rate of change of the longitude of the periastron. Since the precession is $3\pi\mu$ rad/rev, we have

$$\dot{\omega} = 3\pi\frac{\mu}{T}, \tag{6.49}$$

where T is the orbital period of the binary.

We express this formula in terms of the Keplerian parameters of the orbits of the two objects of the binary. Let r_p, r_a denote the periastron and apastron separations, respectively, and recall that $f = r_s/r$. We assume that the μ appearing in (6.49) equals $\tilde{\mu} = \frac{1}{2}(f_p + f_a)$. This assumption is justified as long as the binary has not begun to collapse. Then, using (6.5), we have

$$f_p = \frac{2GM}{r_p c^2}, \quad f_a = \frac{2GM}{r_a c^2} \quad \Rightarrow \mu = \frac{GM}{c^2}\left(\frac{1}{r_p} + \frac{1}{r_a}\right).$$

As mentioned above, the eccentricities of P and of each member of the binary are the same. Hence, if a_1, a_2 are the semi-major axes of S_1 and S_2 respectively, then

$$r_p = (1-e)(a_1 + a_2), \quad r_a = (1+e)(a_1 + a_2),$$

implying that

$$\frac{1}{r_p} + \frac{1}{r_a} = \frac{2}{a}(1-e^2)^{-1}, \tag{6.50}$$

where $a = a_1 + a_2$ is the semi-major axis of P. From Kepler's Third Law (2.26), we have

$$T = 2\pi\sqrt{\frac{a^3}{GM}}, \quad a = (GM)^{1/3}\left(\frac{T}{2\pi}\right)^{2/3}.$$

Substituting all these into (6.49), we obtain

$$\dot{\omega} = 3\pi\frac{2GM}{c^2 T}(GM)^{-1/3}\left(\frac{T}{2\pi}\right)^{-2/3}(1-e^2)^{-1},$$

or

$$\dot{\omega} = 3\frac{(GM)^{2/3}}{c^2(1-e^2)}\left(\frac{T}{2\pi}\right)^{-5/3}. \tag{6.51}$$

This agrees with the post-Keplerian equation for the relativistic advance of the periastron $\dot{\omega}$ given, for example, in [19, 67, 71, 95]. Thus, our model passes this test of GR.

The Hulse-Taylor binary's periastron advance per revolution is known to be

$$d\varphi = 3 \cdot 180 \cdot 6.925 \cdot 10^{-6} = 0.00374°.$$

The period of the binary is $T = 7.7519$ hours, and so the precession is $d\varphi \cdot 365 \times 24/T = 4.226°$ per year, which is the observed one.

Formula (6.51) can be used to compute the total mass M of the system. Combined with the theory-independent mass ratio, this yields the individual masses of the system.

6.4 Orbits in the Strong Field Regime

We take a short break from the tests of GR and consider here both circular and elliptical trajectories in the strong regime of a stationary, static, spherically symmetric body. By strong regime, we mean close enough to the Schwarzschild radius that we can no longer assume that $\tilde{\mu} = \mu$ (see (6.35)). We begin with circular orbits because, as we show now, circular orbits have minimal energy amongst orbits with the same value of $\tilde{\mu}$. Elliptical orbits are handled in Sect. 6.4.2.

Recall that $\tilde{\mu}$ is the average dimensionless potential energy on the trajectory. We show now that for a fixed value of $\tilde{\mu}$, the circular orbit has the least total energy among all elliptical orbits with this value of $\tilde{\mu}$. From (6.37) and (6.35), we have

$$2\mu = 2\tilde{\mu} - 4\tilde{\mu}^2 + f_a f_p = 2\tilde{\mu} - 4\tilde{\mu}^2 + \tilde{\mu}^2(1 - e^2) = 2\tilde{\mu} - 3\tilde{\mu}^2 - \tilde{\mu}^2 e^2.$$

The negative of the product of the roots of of the right-hand side of (6.32) equals $2\mu(\mathcal{E}^2 - 1)$, so

$$2\mu(\mathcal{E}^2 - 1) = -\tilde{\mu}^2(1 - e^2)(1 - 2\tilde{\mu}).$$

Therefore,

$$\mathcal{E}^2 - 1 = -\frac{\tilde{\mu}(1 - e^2)(1 - 2\tilde{\mu})}{2 - 3\tilde{\mu} - \tilde{\mu}e^2}.$$

The graph of $\mathcal{E}^2 - 1$ as a function of the eccentricity e is shown in Fig. 6.5, for $\tilde{\mu} = 5.32497 \times 10^{-8}$.

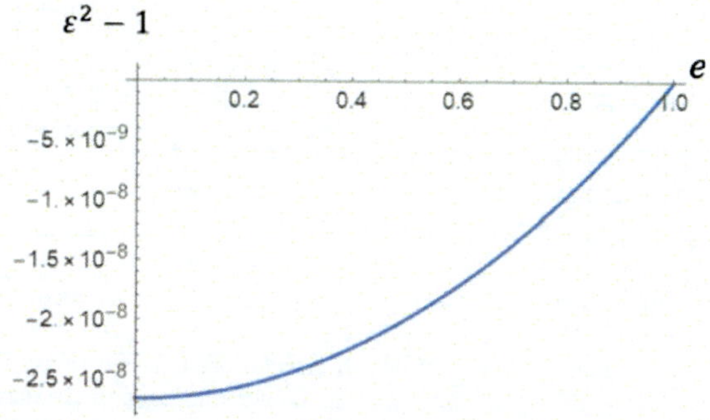

Fig. 6.5 The total energy on an orbit as a function of its eccentricity. The energy $\mathcal{E}^2 - 1$ as a function of the eccentricity e is plotted, for $\tilde{\mu} = 5.32497 \times 10^{-8}$. The energy is minimal when the eccentricity is 0, that is, when the orbit is circular. For the stability of orbits, see Sect. 6.4

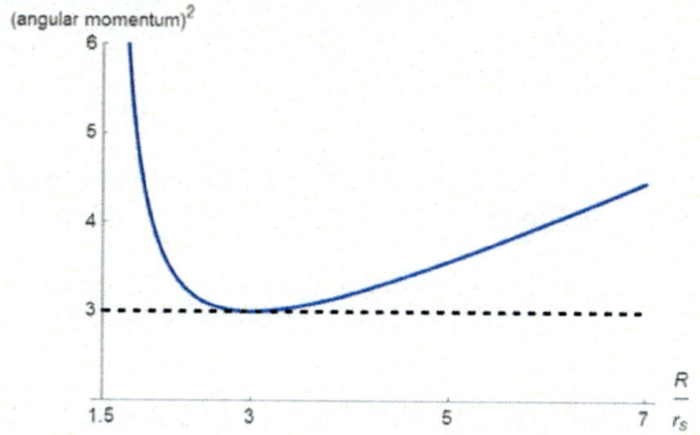

Fig. 6.6 Angular momentum of circular orbits in the strong regime. The square of angular momentum is plotted as a function of the radius of a circular orbit. A circular orbit near $3r_s$ which loses energy will gravitate toward a circular orbit at $3r_s$

6.4.1 Circular Orbits

On a circular orbit, $\rho(\varphi) = R$ for some constant R greater than the Schwarzschild radius. Hence, $R = \rho_p = \rho_a$ and $f_p = f_a = r_s/R$. From (6.34), it follows that

$$2\mu = \frac{2r_s}{R}\left(1 - \frac{2r_s}{R}\right) + \frac{r_s^2}{R^2} = \frac{r_s}{R^2}(2R - 3r_s).$$

Now use (6.31) to obtain

$$\frac{r_s^2}{l^2} = \frac{r_s}{R^2}(2R - 3r_s),$$

or

$$l^2 = \frac{r_s R^2}{2R - 3r_s}. \tag{6.52}$$

From this equation, it follows that a relativistic circular orbit exists if and only if $2R - 3r_s > 0$, or $R > 1.5r_s$. Furthermore, it is easy to check that as a function of R, l^2 has a minimum at $R = 3r_s$ (see Fig. 6.6).

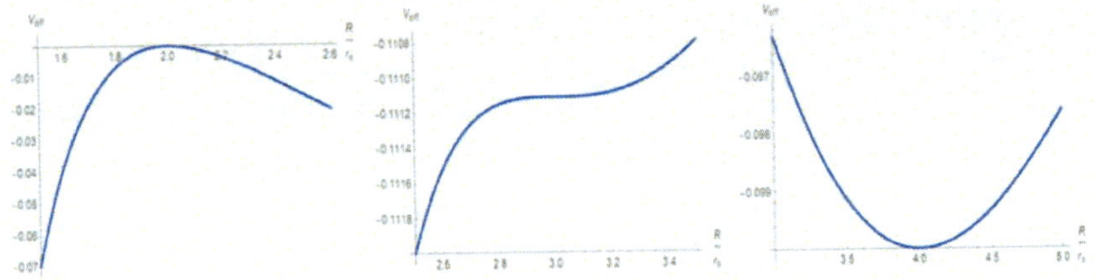

(a) (b) (c)

Fig. 6.7 The stability of circular orbits in the strong regime. The effective potential V_{eff} (6.53), with l^2 given by (6.52), in the neighborhood of circular orbits. **a** The graph of V_{eff} near $R = 2r_s$. Since the effective potential has a maximum at $2r_s$, this is an unstable orbit. **b** The graph of V_{eff} near $R = 3r_s$. The inflection point at $3r_s$ separates the stable orbits ($r > 3r_s$) from the unstable orbits ($1.5r_s < r < 3r_s$). **c** The graph of V_{eff} near $R = 4r_s$. Since the effective potential has a minimum at $4r_s$, this is a stable orbit

Next, we explore the *stability of circular orbits*.To examine what happens in the neighborhood of a circular orbit $r = R$, plot the effective potential

$$V_{\text{eff}} = \frac{l^2}{\rho^2}\left(1 - \frac{r_s}{\rho}\right) - \frac{r_s}{\rho}, \tag{6.53}$$

for r in a neighborhood of R, and with $l^2 = r_s R^2/(2R - 3r_s)$. If V_{eff} has a local minimum at R, then the orbit is stable. If it's a local maximum, then the orbit is unstable. Figure 6.7 shows that circular orbits beyond $3r_s$ are stable, while those between $1.5r_s$ and $3r_s$ are unstable.

Next, we compute the angular velocity of a circular orbit as a function of its radius. From (6.22), it follows that the square of the speed of the object with respect to $\tilde{\tau}$ is $\left(\frac{d\mathbf{x}}{d\tilde{\tau}}\right)^2 = \frac{l^2}{R^2}$. Moreover, since the spatial part of l_μ is radial and the 3D velocity of the orbit is transversal, it follows that $l_\mu \dot{x}^\mu = l_0 \dot{x}^0$. Hence, dividing (4.19) by $d\tilde{\tau}^2$ leads to

$$1 = c^2\left(\frac{dt}{d\tilde{\tau}}\right)^2 - \frac{l^2}{R^2} - \frac{r_s}{R}c^2\left(\frac{dt}{d\tilde{\tau}}\right)^2,$$

implying that

$$\left(\frac{d\tilde{\tau}}{dt}\right)^2 = \frac{c^2\left(1 - \frac{r_s}{R}\right)}{1 + \frac{l^2}{R^2}}.$$

Substitute the expression for l^2 from (6.52) into the above expression to get

$$\left(\frac{d\tilde{\tau}}{dt}\right)^2 = \frac{\frac{c^2}{R}(R - r_s)}{1 + \frac{r_s}{2R-3r_s}} = \frac{\frac{c^2}{R}(R - r_s)}{\frac{2R-2r_s}{2R-3r_s}} = \frac{c^2}{2R}(2R - 3r_s).$$

Let $\Omega = d\varphi/dt$ be the angular velocity of the motion of the object with respect to the lab frame time t. The corresponding angular momentum is $R^2\Omega$ and is equal to $l\frac{d\tilde{\tau}}{dt}$. Using (6.52) and the above expression for $(d\tilde{\tau}/dt)^2$, the square of the lab frame angular momentum is

$$R^4\Omega^2 = l^2\left(\frac{d\tilde{\tau}}{dt}\right)^2 = \frac{c^2}{2}r_sR = GMR.$$

From this, one obtains the angular velocity dependence for relativistic circular orbits:

$$\Omega^2 = \frac{GM}{R^3}, \tag{6.54}$$

which is the *Keplerian* formula (2.26), since $\Omega = 2\pi/T$, where T is the orbital period. Thus, *even in the strong field regime*, the angular velocity of circular orbits is classical!

The square of the linear velocity is

$$R^2\Omega^2 = \frac{GM}{R} = \frac{c^2}{2}\frac{r_s}{R}. \tag{6.55}$$

In order to be less than c^2, we require

$$\frac{GM}{R} < c^2 \Rightarrow \frac{c^2}{2}\frac{r_s}{R} < c^2 \Rightarrow R > \frac{r_s}{2}.$$

Since we already know that $R > 1.5r_s$, the linear speed will be less than c for all circular orbits. In fact, when $R = 1.5r_s$, we have $R^2\Omega^2 = (1/3)c^2$, and so the maximum speed is $\frac{\sqrt{3}}{3}c$. The period T of the orbit is given by Kepler's Laws (see formula (2.26)):

$$T^2 = \frac{4\pi^2}{GM}R^3. \tag{6.56}$$

6.4.2 Elliptical Orbits

Here, we generalize the results of the previous subsection to elliptical orbits. First, we obtain the analog of formula (6.52) for l^2. Let R be the semi-latus rectum of the ellipse. It is known that R is the harmonic mean of the perihelion and the aphelion:

$$R = \frac{2}{\frac{1}{r_p} + \frac{1}{r_a}}.$$

Hence,

$$\tilde{\mu} = \frac{1}{2}\left(\frac{r_s}{r_p} + \frac{r_s}{r_a}\right) = \frac{r_s}{R}.$$

Substituting $\mu = r_s^2/l^2$ (formula (6.31)) and the above expression for $\tilde{\mu}$ into formula (6.35), we have

$$\frac{r_s^2}{l^2} = \frac{2r_s}{R} - \frac{3r_s^2}{R^2} - \frac{e^2 r_s^2}{R^2}. \tag{6.57}$$

This implies that

$$l^2 = \frac{r_s R^2}{2R - (3 + e^2)r_s}. \tag{6.58}$$

Note that this formula reduces to (6.52) for circular orbits ($e = 0$).

Elliptical orbits are not stable, since circular orbits have minimal energy amongst orbits with the same value of $\tilde{\mu}$. The lack of stability can also be seen by graphing the effective potential (6.53), with l^2 defined by (6.58).

The precession of the orbit is computed as follows. From (6.57), we have

$$(3 + e^2)\tilde{\mu}^2 - 2\tilde{\mu} + \frac{r_s^2}{l^2} = 0,$$

which implies that

$$\tilde{\mu} = \frac{1 - \sqrt{1 - 2\mu(3 + e^2)}}{3 + e^2}. \tag{6.59}$$

Expanding the square root in a power series $1 - \mu(3 + e^2) - (1/2)\mu^2(3 + e^2)^2 - \cdots$, we have

$$3\pi\tilde{\mu} = 3\pi\mu + \frac{3\pi}{2}\mu^2(3 + e^2) + \cdots. \tag{6.60}$$

It may be possible to observe the quadratic term for a star such as S2, which has high eccentricity (≈ 0.88) and is close to the black hole at the center of our galaxy (≈ 1.7 Schwarzschild radii at closest approach). When $r \gg r_s$, μ is small, so we ignore all but the first term and

$$3\pi\tilde{\mu} \approx 3\pi\mu,$$

in agreement with (6.38).

6.4.3 Hyperbolic-Like Orbits

In the previous sections, we have considered relativistic circular and elliptical orbits in the gravitation field of a spherically symmetric body. In this section, we treat unbounded orbits, such as those of comets and planetary flybys.

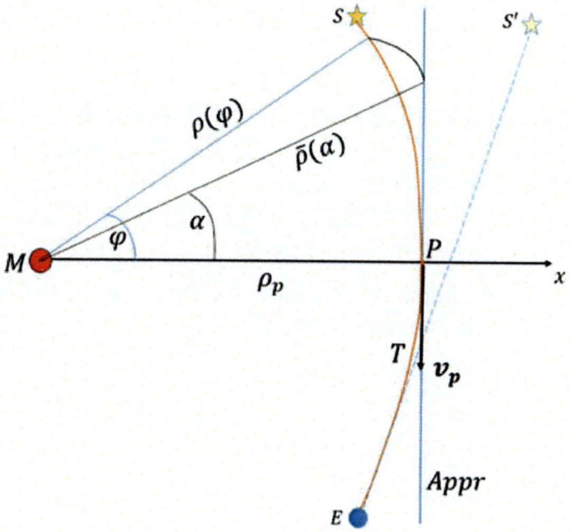

Fig. 6.8 Unbounded orbits and gravitational lensing. M is the source of the field. The actual unbounded trajectory is T (bent, in brown). The straight line approximation is A (in blue). P is the periapsis. Since we will use the approach here also for the bending of light (Sect. 6.5), we have included a star S and its image S' as observed from the Earth E

To compute unbounded orbits, we first note that up to and including Eq. (6.29), the derivation of the worldline of an object in Sect. 6.2 is also valid for unbounded trajectories. For the derivation of unbounded trajectories, the position and the velocity of the moving object at the periapsis P are the initial conditions. Denote by $\mathbf{r}_p$ denote the vector from the source to P, with $|\mathbf{r}_p| = \rho_p$ and by $\mathbf{v}_p = d\mathbf{r}/dt$ its velocity at P, see Fig. 6.8.

Since at P, $\frac{d\rho}{d\varphi} = 0$, Eq. (6.29) implies that

$$\mathcal{E}^2 - 1 = \frac{l^2}{\rho_p^2}\left(1 - \frac{r_s}{\rho_p}\right) - \frac{r_s}{\rho_p}. \tag{6.61}$$

Now rewrite (6.29) as

$$\frac{l^2}{\rho^4}\left(\frac{d\rho}{d\varphi}\right)^2 + \frac{l^2}{\rho^2}\left(1 - \frac{r_s}{\rho}\right) - \frac{r_s}{\rho} = \frac{l^2}{\rho_p^2}\left(1 - \frac{r_s}{\rho_p}\right) - \frac{r_s}{\rho_p}. \tag{6.62}$$

For any angle φ on the trajectory, one may associate an angle $\alpha(\varphi)$ for which $\rho(\varphi) = \bar{\rho}(\alpha)$, that is, the distance from the source to the trajectory equals the distance from the source to the straight-line approximation of the trajectory at the point P. The angle α satisfies $\bar{\rho}(\alpha) = \frac{\rho_p}{\cos\alpha} = \rho(\varphi)$. This suggests the substitution

$$\rho(\varphi) = \frac{\rho_p}{\cos\alpha(\varphi)}, \tag{6.63}$$

which implies that

$$\frac{\rho_p}{\rho} = \cos\alpha, \quad \frac{r_s}{\rho} = \frac{r_s}{\rho_p}\cos\alpha, \quad \frac{d\rho}{d\varphi} = \frac{\rho_p \sin\alpha}{\cos^2\alpha}\frac{d\alpha}{d\varphi}. \tag{6.64}$$

Substituting these into (6.62) and dividing by $\frac{l^2}{\rho_p^2}$ yields

$$\sin^2\alpha\left(\frac{d\alpha}{d\varphi}\right)^2 = 1 - \cos^2\alpha - \frac{r_s}{\rho_p}\left(1-\cos^3\alpha\right) - \frac{r_s}{\rho_p}\frac{\rho_p^2}{l^2}\left(1-\cos\alpha\right).$$

Dividing this equation by $\sin^2\alpha$, using trigonometric identities and $1-\cos^3\alpha = (1-\cos\alpha)(1+\cos\alpha(1+\cos\alpha))$ we obtain

$$\left(\frac{d\alpha}{d\varphi}\right)^2 = 1 - \frac{r_s}{\rho_p}\left(\cos\alpha + \frac{1+\rho_p^2/l^2}{1+\cos\alpha}\right) \tag{6.65}$$

and

$$\frac{d\alpha}{d\varphi} = \sqrt{1 - \frac{r_s}{\rho_p}\left(\cos\alpha + \frac{1+\rho_p^2/l^2}{1+\cos\alpha}\right)}.$$

Thus, the unbounded trajectories in the field of a spherically symmetric source with Schwarzschild radius r_s are described in polar coordinates by a function $\rho(\varphi)$ defined by (6.63), where φ is the angle from the source, measured from the direction to the periapsis P, and $\alpha(\varphi)$ is defined by

$$\alpha(\varphi) = \int_0^{\varphi}\left(1 - \frac{r_s}{\rho_p}\left(\cos\alpha + \frac{1+\rho_p^2/l^2}{1+\cos\alpha}\right)\right)^{-1/2} d\alpha. \tag{6.66}$$

The value of l is determined from the velocity of the moving object at P.

The motion of a massive particle in a gravitation field is independent of the particle's intrinsic properties, even its mass, and depends only on the initial position and velocity. This velocity must belong to the interior of the domain of admissible velocities at this point in the field. For massless particles, on the other hand, the velocity lies on the boundary of this domain, and here, the parameters τ and $\tilde{\tau}$ vanish. To circumvent this problem, we can obtain the parametrization-free trajectory of a massless particle as a limit of the trajectories of massive particles, as the initial velocity approaches the velocity on the boundary. This brings us to the next test of GR.

6.5 Gravitational Lensing

Gravitational lensing is the bending of light rays in the gravitational field of a massive object. We show now that our model predicts the same lensing as GR for a spherically symmetric source.

Light can be considered as photons, which are massless particles. As mentioned at the end of the previous section, we will obtain trajectories of light by taking the limit of unbounded trajectories of massive particles. For massive particles, the trajectory depends only on the initial position and velocity at some point, which was chosen above to be the periapsis P. The velocity at P must belong to the interior of the domain D_P of admissible velocities, defined by (6.13), at the point P. For massless particles, the velocity at P belongs to the boundary ∂D_P of this domain, defined by (6.14).

Our derivation of unbounded trajectories for massive particles uses the parameter $\tilde{\tau}$, defined by (4.19), since only with respect to this parameter is the gravitational force radial, guaranteeing conservation of angular momentum. For massless particles, however, the parameter $\tilde{\tau}$ vanishes.

To get around this problem, we propose to obtain the parametrization-free trajectories of massless particles by taking the limit of parametrization-free trajectories of massive particles, as the initial velocity approaches the velocity on the boundary.

We consider unbounded trajectories with a fixed periapsis P. Since the parametrization-free trajectories of massive particles depends only on the value of l at the point P (which depends on the initial velocity), we have to calculate the limit of l as the velocity approaches the velocity of light at P. Since the velocity at P is transversal to the direction of the field, formula (6.17) implies that $\beta(P) = 1 - l_0^2(P)$. Moreover, for transverse motion, it follows from (4.19) that

$$d\tilde{\tau}^2 = c^2 dt^2 (1 - \beta^2 - l_0^2),$$

implying that

$$l = \rho_p^2 \frac{d\varphi}{d\tilde{\tau}} = \rho_p^2 \frac{d\varphi}{dt}\frac{dt}{d\tilde{\tau}} = \rho_p^2 \frac{d\varphi}{dt} \frac{1}{c\sqrt{1 - \beta^2 - l_0^2}}.$$

Therefore, l approaches ∞ as the velocity approaches the speed of light at P. Thus, for a light trajectory, we have $\frac{\rho_p^2}{l^2} = 0$. Substituting this into Eq. (6.66), we obtain

$$\alpha(\varphi) = \int_0^{\varphi} \left(1 - \frac{r_s}{\rho_p}\left(\cos\alpha + \frac{1}{1+\cos\alpha}\right)\right)^{-1/2} d\alpha. \tag{6.67}$$

Hence, the deflection angle of a light ray traveling from A to B is

$$\delta\phi = \int_{\alpha_A}^{\alpha_B} \left(1 - \frac{r_s}{\rho_p}\left(\cos\alpha + \frac{1}{1+\cos\alpha}\right)\right)^{-1/2} d\alpha - \pi. \tag{6.68}$$

Assuming that the points A and B are very remote from the massive body, the α values α_A, α_B of the points A and B satisfy $\alpha_A \approx -\pi/2, \alpha_B \approx \pi/2$. Moreover, as usual, $r_s/\rho_p \ll 1$, and we use only the first-order term in the power series expansion of the integrand of (6.68). Thus,

$$\delta\phi \approx \int_{-\pi/2}^{\pi/2} \left(1 + \frac{1}{2}\frac{r_s}{\rho_p}\left(\cos\alpha + \frac{1}{1+\cos\alpha}\right)\right) d\alpha - \pi.$$

Since both the integral of $\cos\alpha$ and the integral of $\frac{1}{1+\cos\alpha}$ from $-\pi/2$ to $\pi/2$ are equal to 2, the weak deflection angle becomes

$$\delta\phi \approx \frac{2r_s}{\rho_p} = \frac{4GM}{c^2\rho_p}. \tag{6.69}$$

This is identical to the angle given by Einstein's formula for weak gravitational lensing using GR ([78, 99]). For gravitational lensing in the strong regime, one simply has to take more terms in the power series expansion of the integrand of (6.68).

The bending of light in a gravitational field predicts a change in the position of stars in the sky when they are close to the Sun, see Fig. 6.8. This was predicted by A. Einstein and was verified by Arthur Eddington, Frank Watson Dyson, and their collaborators during the total solar eclipse of May 29, 1919.

6.6 Shapiro Time Delay

As a light ray travels along the unbounded trajectories of the previous section, it experiences the *Shapiro time delay*, the final classical test of GR. By definition, this is the difference in the time it takes light to travel from point A to point B (see Fig. 6.9) in Minkowski space and in a gravitational field. We show now that our model predicts the same delay as GR.

We compute the delay using a limiting process, as in the previous section. We first calculate the time increment on the trajectory for a small change in the angle φ, for motion of massive objects, and then apply the limit as the initial velocity approaches the velocity on the boundary. Using (6.22) and (6.23) for massive objects, we have

$$\frac{cdt}{d\varphi} = \frac{c\dot{t}}{\dot{\varphi}} = \frac{\left(\mathcal{E} + \frac{r_s}{\rho}\dot{\rho}\right)\rho^2}{(1 - r_s/\rho)\, l} = \frac{\mathcal{E}\rho^2}{(1 - r_s/\rho)\, l} + \frac{r_s\rho\dot{\rho}}{(1 - r_s/\rho)\, l}. \tag{6.70}$$

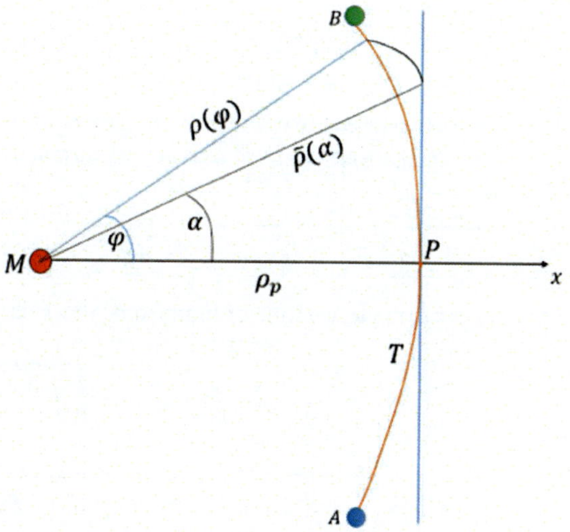

Fig. 6.9 Shapiro Time Delay. M is the source of the field. The actual unbounded trajectory is T (bent, in brown). The vertical blue line is the straight line approximation. P is the periapsis

Since $d\tilde{\tau} = 0$ on light trajectories, the derivatives by $\tilde{\tau}$ tend to infinity when we approach such trajectories. In particular, the angular momentum l tends to infinity. Dividing (6.61) by l^2 and taking the limit for light trajectories, we obtain

$$\frac{\mathcal{E}^2}{l^2} = \frac{1 - r_s/\rho_p}{\rho_p^2} = \frac{1}{b^2},$$

where

$$b = \frac{\rho_p}{\sqrt{1 - r_s/\rho_p}} \tag{6.71}$$

is the *impact factor* on a light trajectory. Next, dividing Eq. (6.26) by l^2, we get

$$\frac{\dot{\rho}^2}{l^2} + \frac{1}{\rho^2}\left(1 - \frac{r_s}{\rho}\right) - \frac{r_s}{l^2 \rho} = \frac{\mathcal{E}^2}{l^2} - \frac{1}{l^2},$$

and in the limit for light trajectories, we obtain

$$\frac{\dot{\rho}^2}{l^2} + \frac{1}{\rho^2}\left(1 - \frac{r_s}{\rho}\right) = \frac{1}{b^2}.$$

This implies that

$$\frac{\dot{\rho}}{l} = \frac{1}{b}\sqrt{1 - \frac{\rho_p^2\,(1 - r_s/\rho)}{\rho^2\,(1 - r_s/\rho_p)}}.$$

Substituting this into (6.70) yields

$$\frac{cdt}{d\varphi} = \frac{\rho^2}{b\,(1 - r_s/\rho)} \left(1 + \frac{r_s}{\rho} \sqrt{1 - \frac{\rho_P^2\,(1 - r_s/\rho)}{\rho^2\,(1 - r_s/\rho_P)}} \right).$$

Since the expression under the square root is less than 1 and $r_s \ll \rho$, we may ignore the square root term and rewrite this formula as

$$\frac{cdt}{d\varphi} = \frac{\rho^2}{b(1 - r_s/\rho)}. \tag{6.72}$$

Hence, the travel time of light from A to P is

$$c(T_P - T_A) = \frac{1}{b} \int_{\varphi_A}^{\pi/2} \frac{\rho^2}{(1 - r_s/\rho)} d\varphi\,. \tag{6.73}$$

To any angle φ on the trajectory, we associate an angle α on the approximation, as shown in Fig. 6.8. Changing variables as in (6.63) and (6.64), the travel time of light from A to P is

$$c(T_P - T_A) = \frac{\rho_P^2}{b} \int_{\alpha_A}^{\pi/2} \frac{1}{\cos^2\alpha(1 - \frac{r_s}{\rho_P}\cos\alpha)} d\alpha\,. \tag{6.74}$$

Here, to make the integral tractable, we have assumed that $d\alpha = d\varphi$. This is justified by formula (6.65) and the fact that $r_s \ll \rho_P$. To calculate the integral, we use partial fractions to decompose the integrand as a sum

$$\frac{1}{\cos^2\alpha(1 - \frac{r_s}{\rho_P}\cos\alpha)} = \frac{1}{\cos^2\alpha} + \frac{r_s/\rho_P}{\cos\alpha} + \frac{r_s^2/\rho_P^2}{1 - \frac{r_s}{\rho_P}\cos\alpha}.$$

The integral of the first summand is $\tan\alpha$, and the integral of $1/\cos\alpha$ is $\ln|\tan\alpha + \sec\alpha|$ (which can be verified by direct differentiation). Since $1 - \frac{r_s}{\rho_P}\cos\alpha \geq 1 - \frac{r_s}{\rho_P}$, the magnitude of the integral of the third summand is less than $\frac{r_s^2\pi}{b(1 - r_s/\rho_P)}$. Since $r_s/b \ll 1$, we can ignore this term.

Assuming that A is very remote from the massive body, we have $\tan\alpha_A = y_A/\rho_P$, where y_A denotes the y coordinate of A, and $\sec\alpha_A = \rho_A/\rho_P \approx y_A/\rho_P$. Thus, from (6.74), the time travel of light from A to P is

$$c(T_P - T_A) \approx \frac{\rho_P}{b}\left(y_A + r_s \ln\frac{2y_A}{\rho_P} \right). \tag{6.75}$$

Assuming that $\frac{\rho_P}{b} \approx 1$, we obtain that the time delay from A to P is $r_s \ln\frac{2y_A}{\rho_P}$. Similarly, the time delay from P to B is $r_s \ln\frac{2|y_B|}{\rho_P}$, and the Shapiro time delay for a signal traveling from A to B and back is

$$2r_s \ln \frac{4y_A|y_B|}{\rho_p^2}, \tag{6.76}$$

which is the known formula for the Shapiro time delay ([78, 99]), confirmed by several experiments.

Therefore, our model passes all of the classical tests of GR. The one remaining test, gravitational waves, will be handled in Sect. 6.9. Note that the final results of the above tests are all independent of the parametrization $\tilde{\tau}$ that we used. In the following sections, we will use the parameter τ, which is independent of the influenced spacetime.

6.7 The Gravitational Field of Multiple Sources

In this section, we consider the gravitational field generated by a finite number of spherically symmetric sources. We assume throughout this section that the sources are stationary, static and non-rotating, although we won't mention this again explicitly. We denote by $x^{(k)}$ the position of the center of source k, and by $r_s^{(k)}$ its Schwarzschild radius.

We define first the action function $L(x,u)$, expressing the influence on the spacetime by the gravitational field of source k. To do this, we perform a translation T on the lab spacetime moving $x^{(k)}$ to the origin, that is, $x' = Tx = x - x^{(k)}$. In coordinates x', the source of the field is at the origin, and the action function (6.1) is

$$L(x', u') = \sqrt{\eta_{\mu\nu}u'^{\mu}u'^{\nu} - (l'_\mu(x')u'^{\mu})^2},$$

where $l'_\mu(x') = \sqrt{\frac{r_s^{(k)}}{|\mathbf{x}'|^3}}(|\mathbf{x}'|, \mathbf{x}')$ by (6.8). Since the action function is Lorentz invariant and the four-vector $u' = u$, we obtain that in the lab frame, the action function $L(x,u)$ is given by (6.1), with

$$l_\mu^{(k)}(x) = \sqrt{\frac{r_s^{(k)}}{|\mathbf{x} - \mathbf{x}^{(k)}|^3}}\left(|\mathbf{x} - \mathbf{x}^{(k)}|, \mathbf{x} - \mathbf{x}^{(k)}\right). \tag{6.77}$$

This implies that

$$(l_0^{(k)})^2 = \frac{r_s^{(k)}}{|\mathbf{x} - \mathbf{x}^{(k)}|}. \tag{6.78}$$

The test particle's Newtonian potential energy in the field generated by the source k, as in Eq. (6.4), is

$$-\frac{1}{2}mc^2(l_0^{(k)}(x))^2 = \Phi^{(k)}(\mathbf{x}), \quad (l_0^{(k)}(x))^2 = -\frac{2\Phi^{(k)}(\mathbf{x})}{mc^2} = \phi^{(k)}(\mathbf{x}), \tag{6.79}$$

where $\Phi^{(k)}(\mathbf{x})$ is the Newtonian potential and

$$\phi^{(k)}(\mathbf{x}) = -\frac{2\Phi^{(k)}(\mathbf{x})}{mc^2} = \frac{r_s^{(k)}}{|\mathbf{x} - \mathbf{x}^{(k)}|} \tag{6.80}$$

is the dimensionless potential energy of the gravitational field generated by the source k.

It is known [59] that the gravitational potential is determined by Poisson's equation

$$\nabla^2 \Phi(\mathbf{x}) = 4\pi G \rho(\mathbf{x}), \tag{6.81}$$

where $\rho(\mathbf{x})$ is the mass density of the sources. This is the field equation in Newtonian gravity, and it depends linearly on the sources of the field. A similar equation holds for $\phi(\mathbf{x})$, which also depends linearly on the sources. Thus, the dimensionless potential energy of our field, generated by a collection of spherically symmetric sources, is

$$\phi(\mathbf{x}) = \sum_k \phi^{(k)}(\mathbf{x}) = \sum_k \frac{r_s^{(k)}}{|\mathbf{x} - \mathbf{x}^{(k)}|}. \tag{6.82}$$

Together with (6.79), this implies that the zero component of the four-potential of the field is related to the zero components of the sources via

$$l_0^2(\mathbf{x}) = \sum_k (l_0^{(k)}(\mathbf{x}))^2 = \sum_k \frac{r_s^{(k)}}{|\mathbf{x} - \mathbf{x}^{(k)}|}. \tag{6.83}$$

This determines l_0 in terms of the sources. It remains to determine the spatial part $\mathbf{l}$ of $l_\mu(x)$. As we have seen, $\mathbf{l}$ is the direction of propagation of the field. As we know from classical physics, the propagation is in the direction of the gradient $\nabla\Phi(\mathbf{x})$ of the potential, which is also the direction of $\nabla\phi(\mathbf{x})$ and $\nabla l_0(\mathbf{x})$. This suggests defining the *four-potential* $l_\mu(\mathbf{x})$ of a field generated by a collection of spherically symmetric sources as

$$l_\mu(\mathbf{x}) = l_0(\mathbf{x}) \left(1, \frac{\nabla\phi(\mathbf{x})}{|\nabla\phi(\mathbf{x})|}\right), \tag{6.84}$$

where $l_0(\mathbf{x})$ and $\phi(\mathbf{x})$ are defined by (6.83) and (6.82), respectively.

We check now that the equation of motion defined by the four-potential $l_u(x)$ has the correct Newtonian limit. Since there are multiple sources k, each one with its own $l_\mu^{(k)}$ and, hence, its own $\tilde{\tau}^{(k)}$, we are forced to use the proper time τ as the parameter on worldlines. Therefore, we use the equation of motion (6.11) for a test object temporarily at rest, with four-velocity $\dot{x} = (1, 0, 0, 0)$. As for a single-source field, also here we have $l \cdot \dot{x} = l_0$ and $(l^\perp)^j = l^j$. Thus, exactly as in (6.12),

$$\ddot{x}^j = -l_0 F_0^j(l) - l_0 F_0^\lambda(l) l_\lambda l^j + l_{0,0} l^j, \quad j = 1, 2, 3, \tag{6.85}$$

where $F^{\alpha}_{\nu}(l) = \eta^{\alpha\lambda}(l_{\nu,\lambda} - l_{\lambda,\nu})$ and $F^{j}_{0}(l) = \eta^{j\lambda}l_{0,\lambda}$. The first term in (6.85) is

$$-l_0 F^j_0(l) = -l_0\eta^{jk}l_{0,k} = -\frac{1}{2}\eta^{jk}(l_0^2)_{,k} = -\frac{1}{2}\eta^{jk}\phi_{,k}.$$

The second term is

$$-l_0 F^{\lambda}_0(l) l_\lambda l = -\frac{1}{2}\left(\nabla\phi \cdot l_0\frac{\nabla\phi}{|\nabla\phi|}\right) l_0\frac{\nabla\phi}{|\nabla\phi|} = -\frac{1}{2}l_0^2\nabla\phi.$$

Since the field is static, the last term vanishes and

$$\ddot{x} = \frac{1}{2}\nabla\phi(1 - l_0^2).$$

This acceleration is with respect to $\tau = ct$ (since the object is temporarily at rest) in our lab frame. But the object's spacetime is influenced by the gravitational field, and its time is t'. Since the time interval dt' in the influenced spacetime is connected to the invariant parameter in the lab frame via (4.50), we have $c^2dt'^2 = d\tau^2 - (l_0 d\tau)^2$, and the 3D acceleration predicted by our model is

$$\mathbf{a}_N = \frac{d^2\mathbf{x}}{dt'^2} = \ddot{x}\frac{d\tau^2}{dt'^2} = \ddot{x}\frac{c^2}{1 - l_0^2} = \frac{c^2}{2}\nabla\phi.$$

Using the connection between ϕ and the gravitational potential Φ, we obtain $\mathbf{a}_N = -\frac{\nabla\Phi}{m}$, which coincides with the acceleration predicted by Newtonian gravity at all points outside the Schwarzschild radius.

6.8 The Gravitational Field of a Moving Source

Consider now the gravitational field generated by a non-rotating, spherically symmetric body of mass M, moving by a worldline $\psi(\tau)$. We wish to compute the acceleration of an object at a spacetime point P (see Figure 3.14). Denote by Q the unique intersection of the worldline of the source and the backward light cone with vertex at P. The relative position null four-vector QP is $r(x) = x - \psi(\tau(x))$. The time $\tau(x)$ is called the retarded time. Only the position of the source at the retarded time has influence at the spacetime position x, and $r(x)$ is the direction of propagation at x. Denote by $w(\tau(x))$ the four-velocity of the source at the retarded time. Using Lorentz covariance of the four-potential $l_\mu(x)$, formula (6.8) can be extended to the gravitational field of a static, spherically symmetric body, with four-velocity $w(\tau)$ with respect to the observer, as

$$l_\nu(x) = \sqrt{\frac{r_s}{(r(x)\cdot w(\tau(x)))^3}}\, r_\nu(x). \qquad (6.86)$$

This is the gravitational analog of the Liénard-Wiechert four-potential (5.16) of a moving charge.

In order to be able to derive the explicit form of the equation of motion (6.11), we have to compute the partial derivatives of $l_\mu(x)$, defined by (6.86). For this, we need the derivatives of the relative position null four-vector $r(x) = x - \psi(\tau(x))$ and the inner product $r \cdot w$. These derivatives were computed in Chap. 3 (formulas (3.88) and (3.90)), and we repeat them here for the reader's convenience. We have

$$r_{\nu,\mu} = \eta_{\mu\nu} - \frac{w_\nu r_\mu}{r \cdot w} \tag{6.87}$$

and

$$(r \cdot w)_{,\mu} = w_\mu + \frac{r_\mu((a \cdot r) - 1)}{r \cdot w}, \tag{6.88}$$

where $a = a(\tau)$ is the four-acceleration of the source at the retarded time.

Using (6.86)–(6.88) yields

$$l_{\nu,\mu} = \sqrt{r_s}\left(\frac{-3}{2}\frac{w_\mu r_\nu + \frac{r_\mu r_\nu((a\cdot r)-1)}{r\cdot w}}{(r\cdot w)^{5/2}} + \frac{\eta_{\mu\nu} - \frac{w_\nu r_\mu}{r\cdot w}}{(r\cdot w)^{3/2}}\right)$$

$$= \sqrt{r_s}\left(\frac{\eta_{\mu\nu}}{(r\cdot w)^{3/2}} - \frac{3}{2}\frac{w_\mu r_\nu}{(r\cdot w)^{5/2}} - \frac{w_\nu r_\mu}{(r\cdot w)^{5/2}} + \frac{3}{2}\frac{r_\mu r_\nu}{(r\cdot w)^{7/2}} - \frac{3}{2}\frac{r_\mu r_\nu(a\cdot r)}{(r\cdot w)^{7/2}}\right).$$

Since $F_{\lambda\nu}(l)$ is antisymmetric in the indices λ and ν, we obtain

$$F_{\lambda\nu}(l) = l_{\nu,\lambda} - l_{\lambda,\nu} = \frac{\sqrt{r_s}}{2}\frac{w_\nu r_\lambda - w_\lambda r_\nu}{(r\cdot w)^{5/2}}.$$

Hence,

$$(l \cdot \dot{x})F^\alpha_\nu(l)\dot{x}^\nu = \frac{r_s}{2}\frac{(r\cdot\dot{x})(w\cdot\dot{x})r^\alpha - (r\cdot\dot{x})^2 w^\alpha}{(r\cdot w)^4}.$$

It follows that this term, the first in the relativistic equation of motion (6.11), falls off at large distances like $1/r^2$. The second term, which is the first term contracted with l and then multiplied by $l^\perp$, falls off like $1/r^3$. In the remaining term $l_{\nu,\mu}\dot{x}^\nu\dot{x}^\mu l^\perp$, the four components which do not contain the acceleration a all fall off like $1/r^2$. Only the component $-\frac{3}{2}\sqrt{r_s}\frac{(r\cdot\dot{x})^2(a\cdot r)}{(r\cdot w)^{7/2}}l^\perp$ falls off like $1/r$. Thus, as in electromagnetism, one obtains two fields: a near field, which falls off like $1/r^2$, and a far field, which falls off like $1/r$. The acceleration due to the far field is

$$\ddot{x}^\alpha = -1.5 r_s \frac{(r\cdot\dot{x})^2(r\cdot a)(r^\alpha - (r\cdot\dot{x})\dot{x}^\alpha)}{(r\cdot w)^5}. \tag{6.89}$$

The gravitational *far field*, like in electromagnetism, depends on the acceleration of the source. However, the electromagnetic far field stems from an antisymmetric tensor, while the gravitational far field comes from the symmetric part of the equation of motion. This concludes the analysis of the gravitational field of a single source moving with respect to an inertial frame.

Static, multiple-source fields were discussed in Sect. 6.7. There, the potentials are not additive, but the total acceleration due to the field is the sum of the accelerations due to the individual sources. Thus, we assume in our model that the acceleration caused by several non-rotating, spherically symmetric, moving sources is also the sum of the accelerations caused by each source. This assumption could be tested, for example, by the gravitational waves emitted during the merging of two neutron stars or black holes. This occurs in the strong field regime, where relativistic effects can be observed. In the next section, we apply our model and this assumption to describe the effect of gravitational waves on a detector. We arrive at the same predictions as in GR. The fact that these predictions have been successfully verified by observation lends credence to our assumption.

6.9 Gravitational Waves

In this section, we describe the action of the far gravitational field of a binary on a gravitational wave detector. We place the detector at the origin of our inertial lab frame K. We choose the x and y axes so that the center of mass of the binary is on the positive x axis and the motion of the binary is in the x, y plane (see Fig. 6.10). We assume that the center of mass of the binary has spatial coordinates $\mathbf{R} = (R_0, 0, 0)$ and is at rest in K. Since all of the physics takes place in the x, y plane, we will use 3D spacetime coordinates. The time coordinate will be $x^0 = ct$, where t is the time in K. Points in the x, y plane are represented by a complex number as $(x, y) = x + iy$. Since only the far field can be detected at such large distances, the acceleration of the mirrors of the detector caused by the far gravitational field of each star is given by (6.89). To apply this formula, we need to know the relative position null vector r between the binary and the detector, as well as the four-velocity and the four-acceleration of the binary in K. Assuming that $R \ll R_0$, we have $r = R_0(1, 1)$. Since the velocity of the detector mirrors is very low, we may assume $\dot{x} = (1, 0, 0)$, which implies that $r \cdot \dot{x} = R_0$. The four-velocity of source j, for $j = 1, 2$, at the retarded time can be written as $w^{(j)} = \frac{1}{\sqrt{1-R_j^2\Omega^2/c^2}}(1, R_j\frac{\Omega}{c}ie^{i\Omega t})$. Ignoring the term $\frac{\Omega^2}{c^2}$ and using the fact that the dot product in the complex plane is $z \cdot u = Re(z\bar{u})$, we obtain $r \cdot w^{(j)} = R_0(1 + R_j\frac{\Omega}{c}\sin(\Omega t))$. Since $w_0^{(j)}$ is constant, $a_0^{(j)} = 0$, and $r \cdot a^{(j)} = -R_0 R_j \frac{\Omega^2}{c^2}\cos(\Omega t)$.

Substituting these expressions and the explicit formula for the Schwarzschild radius into (6.89) yields

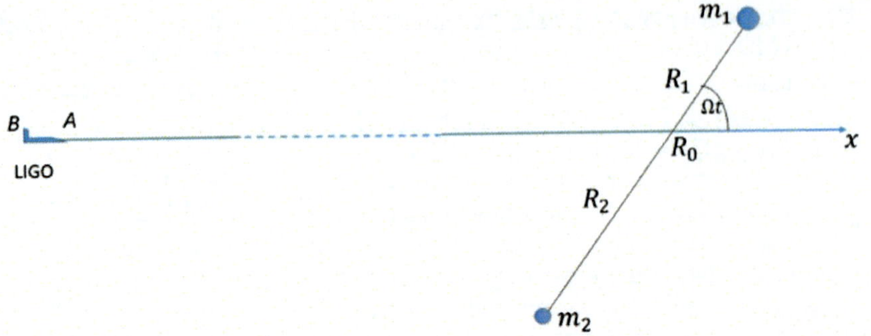

Fig. 6.10 A binary star and a gravitational wave detector. R_0 is the center of mass of the binary. The distance between the stars is R, and the stars are rotating with angular velocity Ω around R_0

$$(\ddot{x}^1)^{(1)} = \frac{1.5 \cdot 2Gm_1}{c^2 R_0} \frac{-m_2 R \frac{\Omega^2}{c^2} \cos(\Omega t)}{M\left(1 + R_1 \frac{\Omega}{c} \sin(\Omega t)\right)^5}$$

and

$$(\ddot{x}^1)^{(2)} = \frac{1.5 \cdot 2Gm_2}{c^2 R_0} \frac{m_1 R \frac{\Omega^2}{c^2} \cos(\Omega t)}{M\left(1 + R_2 \frac{\Omega}{c} \sin(\Omega t)\right)^5}.$$

As mentioned above, we assume that the acceleration of the source caused by the two stars is additive. Thus,

$$\ddot{x}^1 = \frac{3Gm_1 m_2 \Omega^2 R \cos(\Omega t)}{c^4 R_0 M} \left(\left(1 + R_2 \frac{\Omega}{c} \sin(\Omega t)\right)^{-5} - \left(1 + R_1 \frac{\Omega}{c} \sin(\Omega t)\right)^{-5} \right).$$

Approximating $\left(1 + R_j \frac{\Omega}{c} \sin(\Omega t)\right)^{-5} \approx 1 - 5R_j \frac{\Omega}{c} \sin(\Omega t)$, we obtain that the 1-component of the relativistic acceleration is

$$\ddot{x}^1 = \frac{-15Gm_1 m_2 \Omega^3 R^2 \cos(\Omega t) \sin(\Omega t)}{c^5 R_0 M}.$$

Using $\frac{d^2x}{dt^2} = c^2 \ddot{x}^1$ and a trigonometric identity, the acceleration of the mirrors in K is

$$\frac{d^2x}{dt^2} = \frac{-7.5Gm_1 m_2 \Omega^3 R^2 \sin(2\Omega t)}{c^3 R_0 M}. \tag{6.90}$$

Integrating twice by t, we obtain that the displacement of the mirrors in the x direction is

$$\Delta x(t) = \frac{1.875Gm_1 m_2 \Omega R^2 \sin(2\Omega t)}{c^3 R_0 M}.$$

We want to calculate the change of the arm length due to the gravitational wave. Denote by A the mirror at one end of the arm and by B the mirror at the second end. Denote by l the distance between the mirrors. Since the gravitational field propagates with the speed of light, and the distance between the mirrors is negligible with respect to R_0, the displacement $\Delta x_A = \Delta x(t)$ of mirror A at time t equals the displacement $\Delta x_B = \Delta x(t + \Delta t)$, at time $t + \Delta t$, where $\Delta T = l/c$ is the time that takes for the field to propagate from A to B. Thus, the change of distance between the mirrors is

$$\Delta l = \Delta x_B - \Delta x_A = (\Delta x(t))' \frac{l}{c} = \frac{3.75 G m_1 m_2 \Omega^2 R^2 \cos(2\Omega t)}{c^4 R_0 M} l,$$

and the relative distance change is

$$\frac{\Delta l}{l} = \frac{3.75 G m_1 m_2 \Omega^2 R^2 \cos(2\Omega t)}{c^4 R_0 M}. \tag{6.91}$$

This is the same (up to a constant) as obtained by usual methods, see (18.21) of [59].

Chapter 7
Motion of Light and Charges in Isotropic Media

Until this point, we have discussed the relativistic motion of massive objects and massless particles in vacuum, under the influence of electromagnetic and gravitational fields. In this chapter, we will study the motion of photons under the influence of media, as well as the motion of charged particles under the combined influence of an electromagnetic field and media. As in the previous chapters, our approach is based on an action function for the influenced spacetime and the Euler-Lagrange equations and their consequences. We assume that the medium is isotropic, meaning that if the medium is at rest, its influence is the same at each point and independent of the direction in space. We will study the effect of isotropic media, at rest or in uniform motion, on the dynamics of charges and massless particles.

7.1 The Photon Action Function of Rest Media

Consider now the propagation of light in an isotropic medium at rest. This propagation can be considered to be a motion of photons in a spacetime influenced by the medium. Since the charge q of the photon is zero, the parameter $k = 0$ in the simple action function $L(x, u)$ defined by (4.15). Thus, the *action function of photon propagation* in isotropic media is

$$L(x, u) = \sqrt{\eta_{\mu\nu}u^\mu u^\nu - (l_\mu(x)u^\mu)^2}. \tag{7.1}$$

The assumption that the medium is at *rest and isotropic* implies that $l_\mu(x)$ does not depend on the point x. Since the influence is independent of the direction in space, the spatial components of this four-vector must vanish. Thus, we may assume that

Y. Friedman and T. Scarr, *A Novel Approach to Relativistic Dynamics*,
Fundamental Theories of Physics 210,
https://doi.org/10.1007/978-3-031-25214-3_7

$$l_\mu(x) = (l_0, 0, 0, 0), \tag{7.2}$$

where the parameter l_0 depends on the medium. The parameter l_0 may also depend on the frequency of the photon, which is the only intrinsic property of photons distinguishing one from another. The dependence of l_0 on the frequency can be observed in the refraction of light through a glass prism. See Fig. 4.2, in which photons of different frequencies (color) are refracted through different angles. Substituting (7.2) into (7.1), the action function of the photon is

$$L(x, u) = \sqrt{(1 - l_0^2)(u^0)^2 - (u^1)^2 - (u^2)^2 - (u^3)^2}. \tag{7.3}$$

From Chap. 3, we know that the worldline of a massless particle, like a photon, must be null, meaning that the displacement between two events on the worldline is a null vector. We assume that the same property holds for the motion of the photon in media. If we choose the time t to be the parameter σ on a worldline, then with respect to this parameter, the four-velocity is $dx/dt = (c, \mathbf{v})$ where $\mathbf{v}$ is the 3D velocity of the moving object in the medium. The domain D of admissible 3D velocities in the influenced spacetime defined by the action function (7.3) is $D = \{\mathbf{v} \in \mathbb{R}^3 : v^2 \leq c^2(1 - l_0^2)\}$. The velocity $\mathbf{v_0}$ of a photon belongs to the boundary ∂D of D, defined by $v_0^2 = c^2(1 - l_0^2)$. Hence, $v_0 = c\sqrt{1 - l_0^2}$ and $l_0^2 = 1 - \frac{v_0^2}{c^2}$. The *refractive index* n of a medium is defined to be

$$n = \frac{c}{v_0}. \tag{7.4}$$

Note that the speed of light in the medium is always less than the speed of light in vacuum, and may depend on the frequency of the light. In terms of the refractive index, we have

$$l_0 = \sqrt{1 - \frac{1}{n^2}} \quad \text{and} \quad L(x, u) = \sqrt{\frac{1}{n^2}(u^0)^2 - (u^1)^2 - (u^2)^2 - (u^3)^2}. \tag{7.5}$$

This implies that the geometry of the photon's spacetime, influenced by the medium, is the same as that of empty space with the metric

$$d\tilde{s}^2 = v_0^2 dt^2 - dx^2 - dy^2 - dz^2 \ , \ \tilde{\eta}_{\mu\nu} = \operatorname{diag}\left(\left(\frac{v_0}{c}\right)^2, -1, -1, -1\right). \tag{7.6}$$

This is essentially the Minkowski metric (3.37) of flat spacetime, but with c, the speed of light in vacuum, replaced by v_0, the speed of light in the medium.

7.2 The Photon Action Function in Moving Media

In the previous section, we considered the motion of a photon in a medium at rest. Here, we consider photon motion in an isotropic medium moving uniformly with velocity $\mathbf{v}_m$.

The approach to handling this case is to apply Lorentz covariance. Choose an inertial frame in which the medium is at rest. There, apply the techniques of the previous section. Finally, Lorentz transform the results back to the lab frame.

Thus, we assume that the action function of the influenced spacetime of the moving medium has the form (7.1), where the four-covector l_μ in the lab frame is obtained by applying a Lorentz boost with velocity $\mathbf{v}_m$ (a Lorentz transformation of the form (3.18)) to the four-covector l_μ, defined in the comoving frame by (7.2).

To verify the validity of this approach, we revisit the Fizeau experiment, discussed in Sect. 3.3.1 (see Fig. 7.1). By comparing our model's predictions to the results of this experiment, we can check the validity of the assumed form of the action function for moving media.

In the Fizeau experiment, water was made to travel in opposite directions. The speed of light, both with the flow and against the flow, was measured. We will now interpret and analyze this experiment in terms of the action function (7.3).

In an inertial lab frame K', we choose the axes so that the velocity of the water (or other isotropic medium) is a constant $\mathbf{v} = (v_m, 0, 0)$. Let K be an inertial frame in which the water is at rest. Without loss of generality, K and K' are in standard configuration. In K, the rest frame of the water, the action function for a photon is given by (7.3). Since $L(x, u)$ is a Lorentz-invariant scalar, the action function in K' is also given by (7.3), but whose defining covector l'_μ is the Lorentz transform of l_μ.

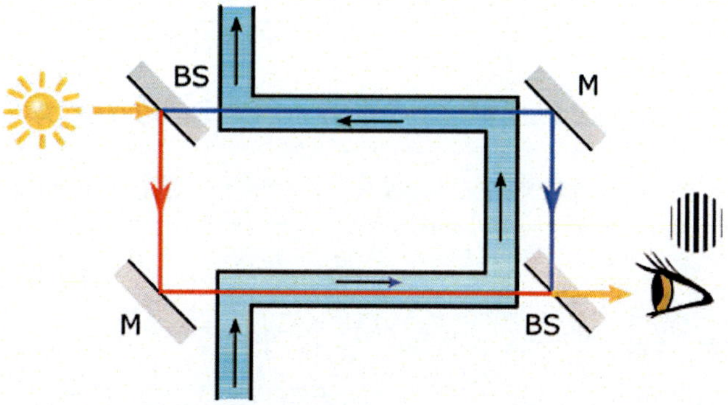

Fig. 7.1 A schematic of the Fizeau experiment. The water flows in two tubes in opposite directions. The light beam from a laser source is split into two beams: one propagates in the direction of water flow, and the other propagates in the direction opposite to the water flow. By observing the interference between the two outcoming beams, it is possible to calculate the speed of light in the moving water. Used with permission from Creative Commons under the license https://creativecommons.org/licenses/by-sa/4.0/deed.en

Recall (see (3.50)) that covectors transform by the *inverse* Lorentz transformation. Hence, $l' = \Lambda l = l_0\gamma(v_m)(1, -v_m/c, 0, 0)$. Let $\beta = v_m/c$. Then $l' = \frac{l_0}{\sqrt{1-\beta^2}}(1, -\beta, 0, 0)$ and

$$(l'_\mu(x)u^\mu)^2 = \frac{l_0^2}{1-\beta^2}((u^0)^2 - 2\beta u^0 u^1 + \beta^2(u^1)^2),$$

implying that $L^2(x, u)$, the square of the action function (7.1) of photon propagation in media, is

$$\left(1 - \frac{l_0^2}{1-\beta^2}\right)(u^0)^2 + \frac{2l_0^2\beta}{1-\beta^2}u^0u^1 - \left(1 + \frac{l_0^2\beta^2}{1-\beta^2}\right)(u^1)^2 - (u^2)^2 - (u^3)^2.$$

To describe the motion of a photon moving in the x direction, we use the lab frame time t' as the parameter of its worldline. With this parameter, the four-velocity in the lab frame is $u = (c, v, 0, 0)$, where v denotes the speed of the photon. Since $L(x, u) = 0$ on the worldline of a photon, we have

$$\left(1 + \frac{l_0^2\beta^2}{1-\beta^2}\right)v^2 - 2\frac{l_0^2\beta}{1-\beta^2}cv - \left(1 - \frac{l_0^2}{1-\beta^2}\right)c^2 = 0,$$

or

$$(1 - \beta^2 + l_0^2\beta^2)v^2 - 2l_0^2\beta cv - (1 - \beta^2 - l_0^2)c^2 = 0.$$

Solving this quadratic for v, we obtain

$$v_{1,2} = \frac{cl_0^2\beta \pm c\sqrt{l_0^4\beta^2 + (1-\beta^2+l_0^2\beta^2)(1-\beta^2-l_0^2)}}{1-\beta^2+l_0^2\beta^2}.$$

Substituting the value of l_0^2 from (7.5) and simplifying, we obtain

$$v_{1,2} = \frac{c\beta\left(1-\frac{1}{n^2}\right) \pm c(1-\beta^2)\frac{1}{n}}{1-\frac{\beta^2}{n^2}} = \frac{\left(c\beta \pm \frac{c}{n}\right)\left(1 \mp \frac{\beta}{n}\right)}{\left(1 \pm \frac{\beta}{n}\right)\left(1 \mp \frac{\beta}{n}\right)}. \tag{7.7}$$

Cancelling and using $c\beta = v_m$ and $\frac{c}{n} = v_0$, we have

$$v_{1,2} = \frac{v_m \pm v_0}{1 \pm \frac{v_m v_0}{c^2}} = v_m \oplus (\pm v_0), \tag{7.8}$$

where $\oplus$ denotes Einstein velocity addition (3.25). This coincides with formula (3.28), which we obtained earlier via a different analysis of the problem.

Since $\beta^2 \ll 1$, we ignore the terms with β^2 in (7.7) and obtain that the velocity of the light in the direction of the motion of the medium is

$$v_+ = v_0 + v_m \left(1 - \frac{1}{n^2}\right),$$

while the velocity in the opposite direction is

$$v_- = -v_0 + v_m \left(1 - \frac{1}{n^2}\right).$$

These results agree with the observed velocities of light in moving water in the Fizeau experiment. This establishes that photon motion in uniformly moving media is modeled correctly by applying Lorentz covariance to the rest case.

7.3 Refraction of Light

Consider now the refraction of a light ray propagating between media separated by a planar surface, which we choose to be the x, y plane. Referring to Fig. 7.2, let n_1 be the refractive index of the upper medium ($z > 0$), and let n_2 be the refractive index of the lower medium ($z < 0$). We choose the x axis so that the incoming ray is in the plane $y = 0$.

Using the action function $L(x, u)$, defined by (7.5), the momenta (2.31) along a worldline $x(\sigma)$ are

$$p_0 = \frac{1}{L}\frac{1}{n^2}\frac{dx^0}{d\sigma} \quad \text{and} \quad p_j = \frac{1}{L}\frac{dx^j}{d\sigma}, \quad j = 1, 2, 3.$$

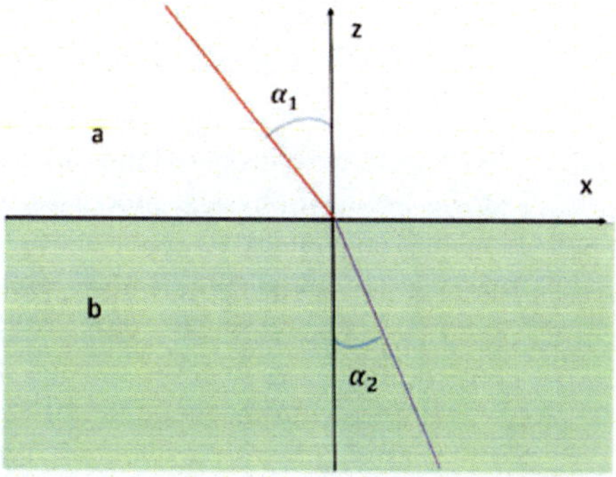

Fig. 7.2 Refraction of light between two media. The upper medium a and the lower medium b are separated by a planar surface containing the x axis

Thus,

$$\frac{p_j}{p_0} = \frac{n^2 dx^j/d\sigma}{dx^0/d\sigma}.$$

The function $L(x, u)$ for our problem *is* dependent on $x^3 = z$, through $n(z)$, but is independent of x^0, x^1 and x^2. Thus, from (2.33), it follows that on a stationary worldline, $\frac{p_1}{p_0}$ is conserved.

Note that p_0 and p_j are not defined on the worldline of a photon, since there $L = 0$. As a result, certain quantities are infinite on a photon's worldline. Nevertheless, there are certain well-defined *ratios* of quantities along worldlines approaching the photon's worldline. Moreover, the limit of these ratios is finite and thus well defined on the photon's worldline as well. This approach was used earlier, for the gravitational lensing in Sect. 6.5 and for the Shapiro time delay in Sect. 6.6. Since the photon worldline is stationary, if we choose the parameter $\sigma = t$, then

$$\frac{p_1}{p_0} = \frac{v_x n^2}{c}$$

is conserved on this worldline.

For the upper medium, the x component of the velocity $v_x = \frac{c}{n_1} \sin \alpha_1$, where α_1 is the incidence angle of the incoming ray. Thus

$$\frac{p_1}{p_0} = n_1 \sin \alpha_1.$$

Similarly, for the lower medium, $\frac{p_1}{p_0} = n_2 \sin \alpha_2$. Since $\frac{p_1}{p_0}$ is conserved on a stationary worldline, we obtain

$$n_1 \sin \alpha_1 = n_2 \sin \alpha_2.$$

This is Snell's Law for light refraction. Conclusion: the refraction of light is an expression of the conservation of the ratio of a photon's momenta.

7.4 Motion of a Charge in an Isotropic Medium at Rest

In Chap. 5, we considered motion of charges in an electromagnetic field in vacuum. However, most physical processes occur in media, where there are bound electric charges and magnetic dipoles which contribute to the electromagnetic field. We assume that when there is no external electromagnetic field, the medium is neutral and will not affect moving charges. However, when there is an electromagnetic field present, the bound electric charges and magnetic dipoles in the medium produce a secondary electromagnetic field. As a result, the motion of a charge in media under a primary field, with a source defined by free current densities j, will be affected not only by the primary field, but also from the secondary field emerging from charges and dipoles in the media. Since the quantity of bound charges and dipoles is extremely large, and the field generated by them depends on the primary field, we will not consider the medium to be an electromagnetic source. Rather, the influence of the medium will be expressed in the covector $l_\mu(x)$. Since the medium is assumed to be homogeneous, l_μ will not depend on x.

In this section, we restrict ourselves to isotropic media at rest. This implies that the effect of the medium is the same at each spatial point and is independent of the direction. Consider the motion of a charge q of mass m in an electromagnetic field in a rest isotropic medium. The spacetime of the charge is influenced by both the electromagnetic field and the medium. The electromagnetic field is described by the four-potential $A_\mu(x)$, as in Chap. 5. The effect of the medium is expressed by $l = (l_0, 0, 0, 0)$, with l_0 defined by (7.5). This ensures that the velocity of massive, non-radiating objects in the medium remains below the speed of light in the medium. Thus, the action function for a charge in an isotropic medium is

$$L(x, u) = \sqrt{\tilde{\eta}_{\mu\nu} u^\mu u^\nu} + k A_\nu(x) u^\nu , \tag{7.9}$$

which is (5.1), with $\eta_{\mu\nu}$ replaced by $\tilde{\eta}_{\mu\nu}$, defined by (7.6). Note that this action function is valid only if the speed of the moving charge is less than the speed v_0 of light in the medium.

To derive the equation of motion, we will use the parameter $\tilde{\tau}$ defined by (4.19), that is

$$d\tilde{\tau}^2 = \tilde{\eta}_{\mu\nu} dx^\mu dx^\nu .$$

From (4.52), the $\tilde{\tau}$ derivatives of the momenta are

$$\dot{p}_\lambda = \eta_{\lambda\mu}\ddot{x}^\mu - l_\mu \ddot{x}^\mu l_\lambda + k A_{\lambda,\nu}\dot{x}^\nu . \tag{7.10}$$

The four-force is

$$\frac{\partial L(x, u)}{\partial x^\lambda} = k A_{\nu,\lambda}\dot{x}^\nu . \tag{7.11}$$

By the Euler-Lagrange equations, the equation of motion is

$$\ddot{x}^\alpha - l_0 \ddot{x}^0 l^\alpha = k F^\alpha_\nu(A)\dot{x}^\nu . \tag{7.12}$$

Hence,

$$(1 - l_0^2)\ddot{x}^0 = k F^0_\nu(A)\dot{x}^\nu \quad \text{and} \quad \ddot{x}^j = k F^j_\nu(A)\dot{x}^\nu . \tag{7.13}$$

This is the relativistic equation of motion in an electromagnetic field in media, where the differentiation is by $\tilde{\tau}$ and $k = \frac{q}{mc^2}$.

If the speed of the charge entering the medium is larger than v_0, then $\tilde{\eta}_{\mu\nu} u^\mu u^\nu < 0$. In this case, the charge radiates the so-called *Cherenkov radiation*. This radiation changes the influenced spacetime of the charge and should also change the action function.

Chapter 8
Spin and Complexified Minkowski Spacetime

In the previous chapters, we have considered the relativistic evolution of point-like objects, without considering their internal rotations or spin. Here, we begin to take spin into account. In Newtonian dynamics, the motion of a (macroscopic) object is decomposed into the motion of the object's center of mass and its rotation about this center. The rotation about the center of mass will be called the object's *spin*. The first step is to broaden our model to incorporate a particle's spin. We point out that in the literature, spin has been considered a purely quantum property with no classical analog.

Here, we propose a relativistic description of spin. The four-velocity w of a particle is a four-vector in Minkowski space. In order to incorporate a particle's spin s, we introduce a complexified Minkowski space and a complex four-vector $\varrho = w + is$. We will see that for particles, the real and complex parts of ϱ are related in a certain way. Using this representation of particles, we derive the formulas for the transition probabilities of spin-1/2 particles. These transition probabilities agree with those of quantum mechanics. We also show that the predictions of our model agree with experimental observations involving the spin of spin-1/2 particles.

Using our representation of spin, we extend the equation of motion (4.27) for charged particles in an electromagnetic field to incorporate spin. The evolution of spin so obtained agrees with the known results in the literature. We reveal the connection between a particle's angular and magnetic momentum and show that the Landé factor must be $g = 2$.

Y. Friedman and T. Scarr, *A Novel Approach to Relativistic Dynamics*,
Fundamental Theories of Physics 210,
https://doi.org/10.1007/978-3-031-25214-3_8

8.1 History of the Spin of Particles

The usual definition of *spin* is an intrinsic form of angular momentum carried by elementary particles. The spin is due to the rotation of the particle. If, in addition, the particle is charged, then its rotation generates a *magnetic moment*. It is known from classical electrodynamics that a rotating, electrically charged body creates a magnetic dipole, with magnetic poles of equal magnitude but opposite polarity. The magnetic strength and orientation of this dipole is called the *magnetic moment* of the particle. In an external magnetic field, the magnetic moment experiences a torque, depending on its orientation with respect to the field.

The existence of electron spin angular momentum was demonstrated by the Stern-Gerlach experiment [50], see Fig. 8.1. In this experiment, a beam of silver atoms is passed through an inhomogeneous magnetic field. Upon exiting the field, the beam splits into two, with half of the atoms being deflected up and the other half being deflected down. This reveals that the evolution of an atom in the field depends on the atom's magnetic momentum. In this case, the magnetic momentum is due to a single electron in the outermost orbit of the silver atom. Moreover, the electron's spin is *quantized*. There are only two possible values, not a continuum of values as predicted by classical physics.

Another indication of the existence of spin is the *anomalous Zeeman effect*. Spectral lines of an atom result from a transition between stable states of the atom and reveal information about the energy difference between these states. The Zeeman effect is the splitting of spectral lines in the presence of an external magnetic field. The main splitting is traditionally called the *normal effect*, while there is an additional splitting, called the *anomalous Zeeman effect*, caused by a nonzero difference in spin between the initial and final states. An additional shift in the spectral lines comes from the interaction of the spin with the particle's orbital angular momentum. This shift defines the *fine-structure interval* and is yet another indication of the existence of spin. Many other experiments prove the existence of both an electron's angular and magnetic momenta in various physical phenomena.

The relativistic quantum theory of electron spin was developed by Dirac [21]. In this theory, the spin state is represented by *bispinors*, a formal mathematical construct invented for this problem. In Dirac's approach, one uses a spin-1/2 representation of the Lorentz group on the bispinors. There are alternative approaches as well to the relativization of spin. See, for example, [83] and [60]. It is claimed [17] that spin has no classical analog.

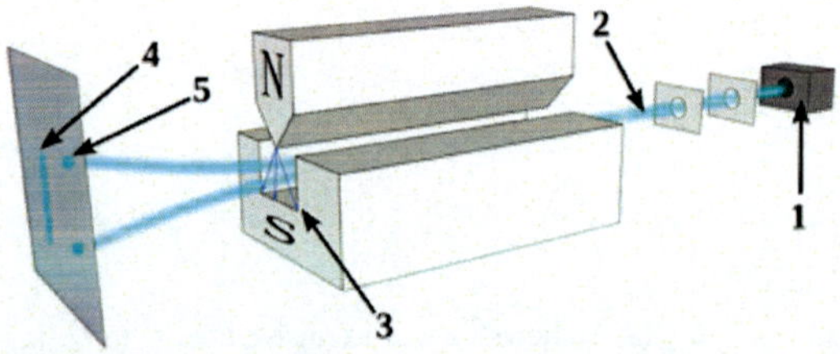

Fig. 8.1 The Stern-Gerlach experiment. Used with permission from Creative Commons under the license https://creativecommons.org/licenses/by-sa/4.0/deed.en

Electron spin has been *partially* treated by classical electrodynamics [61] and special relativity. Uhlenbeck and Goudsmit [93] hypothesized that an electron's magnetic momentum μ and its angular momentum $\tilde{\mathbf{s}}$ are related by

$$\mu = \frac{gq}{2mc}\tilde{\mathbf{s}}, \tag{8.1}$$

where q and m are, respectively, the charge and mass of the electron, c is the speed of light in vacuum, and g is the *Landé factor*.

The classical value for g is 1. Uhlenbeck and Goudsmit observed that the anomalous Zeeman effect could be explained if we assume that $g = 2$. However, under this assumption, the observed fine structure interval is one half of that predicted. Thomas [92] has shown that if one uses relativistic kinematics to describe the precession of the frame comoving to the electron, then $g = 2$ explains both the anomalous Zeeman effect and the observed fine structure interval. This is an indication that spin should be treated within relativity theory. Partial resolution of this problem is given in [61], Sect. 11.11 in the derivation of the relativistic Bargmann, Michel and Telegdi (BMT) equation of motion of spin in an external electromagnetic field. The spin state of a particle is described by a four-vector $s = s^\mu$ which satisfies $s \cdot w = 0$, where w is the four-velocity of the particle. What is missing in this approach is the definition of a particle, which we define below. Our approach is a continuation of [47] and [46].

8.2 The State Space of an Extended Object Moving Uniformly

In this section, we discuss the meaning of the *state* of an object, that is, a *complete, minimal* description of the object at a given time. *Completeness* means that the state at a given time, together with the laws of Nature, are enough to define uniquely the state of the system in the future. A description is *minimal* if omitting some information from the state implies loss of this uniqueness.

In the previous chapters, the motion of an object or particle is described by a worldline $x(\tau)$, denoting the spacetime position of the object at time τ, for some parameter τ. The corresponding equation of motion is a second-order differential equation in τ. It is well known from the theory of differential equations that such an equation has a unique solution for a given initial position x and four-velocity $\dot{x}$. Thus, in Chaps. 4–6, one may take the pair $(x(\tau), \dot{x}(\tau))$ as the state of the object. Indeed, all of our equations of motion involve no more than x and $\dot{x}$. Note also that the action function (4.15) is a function of x, u where $\dot{x}$ is substituted for u.

Moreover, a state $(x(\tau), \dot{x}(\tau))$ is comprised of a spacetime position $x(\tau)$ in Minkowski space M, and a four-velocity $\dot{x}(\tau)$, which is a four-vector in $\tilde{M}$. Thus, the state belongs to the algebraic sum $M \oplus \tilde{M}$. Minkowski space is "big enough" to describe the state for the relativistic motion of particles (without considering their spin).

In this chapter, we expand our horizons and consider the motion of particles and rigid bodies which also spin (rotate). To achieve this, we have to extend our description of the state of a moving object. In addition to the position of its center or center of mass at any given time, we will have to define also its *angular position* at that time. We consider first the state of objects moving uniformly and extend later to non-uniform motion. As a first step, we will consider the problem in the inertial frame comoving with the object. The next step is to transfer this description of the spin state to the lab frame using Lorentz covariance. Let $K'(\tau)$, with basis e'_μ, be the inertial frame comoving with the object. In this frame, the four-velocity of the object with worldline parameterized by the parameter τ is $w' = e'_0$. To define the angular position of the object, attach a spatial frame $\tilde{K}(\tau)$ to the center of the object with $\tilde{K}(0) = K$. The angular position of the object at time τ is defined by a rotation matrix $\Phi(\tau)$, transforming the spatial basis of K' to $\tilde{K}$ at time τ. The angular velocity matrix L' is defined to be the derivative by τ of the rotation matrix $\Phi(\tau)$.

A rotation matrix has a norm one eigenvector $\mathbf{b}'$, with eigenvalue 1, corresponding to the axis of rotation. The rotation is by some angle φ with respect to this axis, and depends on the choice of orientation of the space. The rotation angle φ becomes $-\varphi$ with a change of orientation. Thus, we can define the angular position of the object by the unit-free pseudovector $\phi' = \varphi(0, \mathbf{b}')$. This is a purely spatial pseudovector in K', meaning $w' \cdot \phi' = 0$. Since the derivative L' of a one-parameter family of rotation matrices $\Phi(\tau)$ is an antisymmetric 3×3 matrix, the angular velocity may also be identified with a purely spatial pseudovector $s' = |s'|(0, \mathbf{b}')$ in direction of the rotation axis in K' such that $L'\mathbf{v} = \mathbf{v} \times \mathbf{b}'$ for all $\mathbf{v} \in \mathbb{R}^3$. The magnitude of s' is

$$|s'| = \frac{\omega}{c} \ , \quad s' = \frac{\omega}{c}(0, \mathbf{b}'), \tag{8.2}$$

where ω is the angular velocity.

In order to be able to combine a pseudovector with a position vector, which does not depend on the orientation, we introduce a pseudoscalar i which also changes sign with a change of orientation. Thus, the quantities $i\phi'$ and is' will be independent of the choice of orientation. A similar idea is used in electromagnetism in the definition of the Faraday vector $\mathbf{F} = \mathbf{E} + ic\mathbf{B}$. This helps to obtain explicit, analytic solutions [34, 35] of the motion of charges in an electromagnetic field. However, to be able to combine the position x' and the angular position $i\phi'$, they must have the same dimensions. Since x' has dimensions of length and $i\phi'$ is unit free, the extended spacetime position should be $x' + iR\phi'$, for some constant R, depending on the object, with dimensions of length. We denote the full position as $x'_c = x' + iR\phi'$. Similarly, to be able to combine the unit-free four-velocity and the angular velocity, we will multiply the angular velocity by R and define the *complexified four-velocity* as $u'_c = w' + iRs'$.

For a spherically symmetric object of radius R, the quantity $R\phi'$ is the arc length on the object's surface in the plane of rotation. For particles, we define R as follows. We assume, as did Compton [15], that the wavelength of a particle is equal to the wavelength of a photon whose energy is the same as the rest energy of that particle. Using Planck's law $E = \hbar\omega = mc^2$, we have

$$R = \frac{c}{\omega} = \frac{\hbar}{mc}, \tag{8.3}$$

which has dimensions of length. For example, the above formula gives the Compton radius of the electron to be approximately $R \approx 3 \times 10^{-13}$ m. Thus, for a particle,

$$Rs' = (0, \mathbf{b}') \ , \ (Rs')^2 = -1, \tag{8.4}$$

where $\mathbf{b}'$ is the direction of the axis of rotation.

To describe the position and angular position of the object in the lab frame, we perform a Lorentz transformation Λ, defined in Chap. 3, from the comoving frame to the lab frame. We assume that this transformation extends linearly to the complex part, so that $x_c = \Lambda x'_c = \Lambda x' + i\Lambda R\phi' = x + iR\phi$. Since the parameter τ is the same in both the lab frame and the comoving frame, this suggests that the complex velocity in the lab frame will be $\Lambda u'_c = \Lambda w' + i\Lambda Rs' = w + iRs$. The real part of the complex four-velocity is the usual four-velocity, while the imaginary part is spatial and orthogonal to the four-velocity, since Λ preserves the Minkowski inner product.

8.3 Complexified Minkowski Space as the State Space of an Extended Object

In light of the previous considerations, we have to extend Minkowski space M to *complexified Minkowski space* M_c, which we define as

$$M_c = M + iM. \tag{8.5}$$

Worldlines $x_c(\tau) = x(\tau) + iR\phi(\tau)$ are now complex valued and will be parameterized by the proper time τ (3.38). The *complexified four-velocity* $dx_c(\tau)/d\tau$ is a complex, dimensionless four-vector

$$u_c(\tau) = w(\tau) + iRs(\tau) = \frac{dx(\tau)}{d\tau} + iR\frac{d\phi(\tau)}{d\tau} \tag{8.6}$$

in *complexified Minkowski four-velocity space*

$$\tilde{M}_c = \tilde{M} + i\tilde{M}. \tag{8.7}$$

The real part of the complexified four-velocity $w(\tau)$ is the four-velocity of the object, while the imaginary part is the pseudovector $Rs(\tau)$, which is proportional to the angular velocity. Note that $s(\tau) \cdot w(\tau) = 0$. The direction of $s(\tau)$ is the direction of the axis of rotation.

We extend the Minkowski inner product (3.51) to $\tilde{M}_c$, as follows. Denote the natural basis of $\tilde{M}_c$ by e_0, e_1, e_2, e_3. Any element $x \in \tilde{M}_c$ can be written as $x = x^\mu e_\mu$, for unique $x^\mu \in \mathbb{C}$. The *complex inner product* of two elements $x, y \in \tilde{M}_c$ is defined by

$$x \star y = \eta_{\mu\nu} x^\mu y^\nu, \quad \eta = diag(1, -1, -1, -1). \tag{8.8}$$

Note that this inner product is linear in both terms, symmetric in x and y, and complex valued. The norm of a vector $x \in \tilde{M}_c$ is defined by

$$x^2 = \eta_{\mu\nu} x^\mu x^\nu = x \star x.$$

The *complex conjugate* of x is $\overline{x} = \overline{x}^\mu e_\mu$.

Note that even though the complex inner product is complex valued, the complex four-velocity has positive norm. Since w is timelike, we have $w^2 > 0$. Since s is spatial and orthogonal to w, we have $s^2 < 0$ and $w \star s = 0$. Hence,

$$u_c^2 = (w + iRs)^2 = w^2 + (iRs)^2 = w^2 - (Rs)^2 > 0. \tag{8.9}$$

To justify our definitions of complexified Minkowski space and the inner product, we show that they predict properly the motion of extended objects in free space. The *action function in complexified space* is

$$L(x_c, u_c) = \sqrt{\eta_{\mu\nu} u_c^\mu u_c^\nu}, \tag{8.10}$$

with u_c defined by (8.6). This action function is a natural extension of (3.57). Note that from (8.9), it follows that the action function is well defined.

The momenta are

$$p_\mu = \frac{\eta_{\mu\nu} w^\nu}{\sqrt{\eta_{\mu\nu}(w + iRs)^\mu (w + iRs)^\nu}} + i \frac{\eta_{\mu\nu} Rs^\nu}{\sqrt{\eta_{\mu\nu}(w + iRs)^\mu (w + iRs)^\nu}}. \tag{8.11}$$

Since $L(x, u)$ does not depend on x^μ, the real and the imaginary parts of all momenta are conserved. Letting L denote the square root in the denominator of the momenta, we see that

$$\text{Re}\, p_0 = \frac{w^0}{L}, \quad \text{Re}\, p_j = -\frac{w^j}{L}$$

are conserved. From the definition of the four-velocity, $w^0 = c\frac{dt}{d\tau}$ and $w^j = \frac{dx^j}{d\tau}$. Let $\mathbf{v}$ denote the 3D velocity, with $v^j = dx^j/dt$. Then

$$v^j = \frac{dx^j}{dt} = c\frac{w^j}{w^0} = -c\frac{\operatorname{Re} p_j}{\operatorname{Re} p_0}$$

is conserved, implying that the motion of the center of mass is uniform, as it should be. In particular, the comoving frame is the same for all τ.

Conservation of the imaginary part implies that

$$\operatorname{Im} p_\mu = R\frac{\eta_{\mu\nu}s^\nu}{L}$$

is conserved and also $\frac{s}{L}$ is conserved. Using (8.6), this implies that the angular velocity with respect to time t

$$\frac{d\phi}{dt} = \frac{d\phi}{d\tau}\frac{d\tau}{dt} = \frac{cs}{w^0}.$$

is conserved, as it should be. Thus, in complexified Minkowski space, a stationary worldline has constant velocity and constant angular velocity, in agreement with our experience.

8.4 The Representation of the Spin of an Electron

When we consider the state of the spin of an electron and its transition from one state to another, we ignore the electron's position and linear motion and consider its state to be defined solely by its complexified velocity u_c. Moreover, we may assume that its linear velocity is negligible with respect to the speed of light c, implying that we may assume that $w = e_0$.

Based on the Stern-Gerlach (SG) experiment and the results of experiments involving a sequence of SG apparatuses, we obtain here a condition on the relationship between the real and the imaginary parts of the complex four-velocity of an electron. See Fig. 8.2.

In setup (a), a stream of electrons passes an SG apparatus and is split into two beams, one going up, with state ϱ_+, and one going down, with state ϱ_-. After the beams pass this first apparatus, the ϱ_- electrons are artificially blocked, so that only the ϱ_+ electrons are allowed to pass through the second apparatus. All of the electrons passing though the second apparatus go up, showing that their state is ϱ_+.

Mathematically, we represent the process of sending the electrons through one SG apparatus and blocking the ϱ_- beam by a map P_+, indicating that all of the electrons which exit the apparatus have state ϱ_+. Similarly, if one blocks the exiting ϱ_+ electrons, we obtain a map P_-. In part (a) of the figure, we see that $P_+^2 = P_+$. In part (b), we see that $P_- P_+ = 0$.

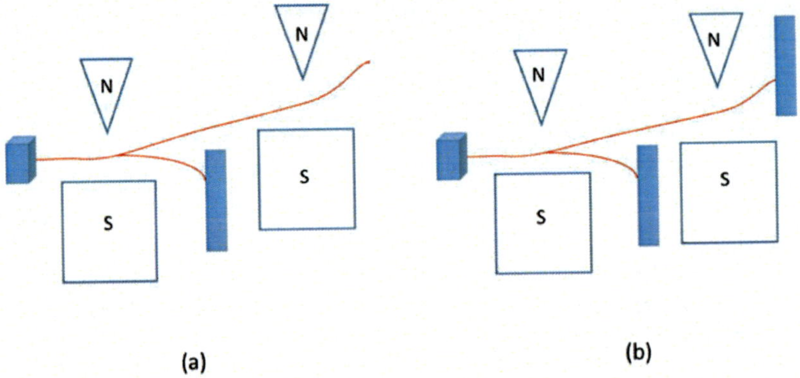

Fig. 8.2 An electron beam from the furnace passes a sequence of two SG apparatuses. **a** After the first SG, the beam ϱ_- is artificially blocked. The beam ϱ_+ passes the second SG, with all electrons going up. **b** After the first SG, the beam ϱ_- is blocked. The beam ϱ_+ passes the second SG and then ϱ_+ is blocked. No electrons pass through

It is known and easy to verify that the operator P_+ defined by

$$P_+(\varrho) = \frac{\varrho_+ \star \varrho}{\varrho_+^2}\varrho_+ \tag{8.12}$$

is a projection onto particles with state ϱ_+. Similarly, $P_-(\varrho) = \frac{\varrho_- \star \varrho}{\varrho_-^2}\varrho_-$ is a projection onto particles with state ϱ_-.

The equation $P_- P_+ = 0$ implies that

$$P_- P_+(\varrho) = P_- \frac{\varrho_+ \star \varrho}{\varrho_+^2}\varrho_+ = \frac{\varrho_+ \star \varrho}{\varrho_+^2}\frac{\varrho_- \star \varrho_+}{\varrho_-^2}\varrho_- = 0$$

for any ϱ. This implies that $\varrho_- \star \varrho_+ = 0$. If we represent the spin up state by $\varrho_+ = w + iRs$, with $w \cdot s = 0$, then the spin down state is $\varrho_- = w - iRs$. Thus, $\varrho_- \star \varrho_+ = (w + iRs) \star (w - iRs) = w^2 + (Rs)^2 = 0$, which implies that for the spin state of an electron, the angular velocity must satisfy

$$(Rs)^2 = -1. \tag{8.13}$$

This also follows from (8.4) by use of Compton's assumption.

As mentioned in [7], in non-relativistic quantum mechanics, the spin state of a particle has a definite direction (more precisely, there exists a direction in which the spin component has definite value). The angular part of the spin state of such a particle is defined by a unit vector $\mathbf{s} \in \mathbf{R}^3$ representing the direction of the spin, with the meaning that the rotation is in the plane perpendicular to $\mathbf{s}$. This vector is called the *Bloch vector* [7], and the collection of all Bloch vectors is called the *Bloch sphere*. The Bloch sphere is a unit sphere in $\mathbb{R}^3$, representing the pure state space of a two-level quantum mechanical system, with antipodal points corresponding to a pair of mutually orthogonal state vectors. It is used in quantum mechanics and also in quantum computing.

Thus, we have shown that in the comoving frame, the spin state of an electron, defined by its complexified four-velocity, is of the form $\varrho' = e_0 + is'$, with $(s')^2 = -1$. In the lab frame K, the *spin state of an electron*, or any other spin-1/2 particle, is

$$\varrho = w + iRs, \quad w^2 = 1, \quad w \cdot s = 0, \quad (Rs)^2 = -1. \tag{8.14}$$

The four-vector s describing the angular part of the spin state is called the *4D spin vector* in [61]. It is also known as the *Pauli-Lubansky spin vector*. Note that the relativistic spin state ϱ in our representation is a well-defined, physically meaningful complex four-vector and not an abstract bispinor, as in the Dirac formulation. Moreover, we can consider the spin state as the total velocity of a rotating object, where the real part is the four-velocity and the imaginary part is its dimensionless angular velocity.

Note that $\varrho^2 = w^2 - (Rs)^2 = 2$. As a result, we introduce a *normalized spin state*

$$\hat{\varrho} = \frac{1}{\sqrt{2}}\varrho = \frac{1}{\sqrt{2}}(w + iRs). \tag{8.15}$$

8.5 Transition Probabilities of Spin States and Bell's Inequality

The *transition probability* between two spin states ϱ_1 and ϱ_2 is the proportion of particles in state ϱ_1 that passes the second SG apparatus to reach state ϱ_2. Since the transition probability should be independent of the observer's reference frame, we will work in the frame comoving with the particles. To determine the transition probability between two spin states ϱ_1 and ϱ_2, let $\mathbf{a}$ denote the spin direction of ϱ_1, and let $\mathbf{b}$ denote the spin direction of ϱ_2. Since $|\mathbf{a}| = |\mathbf{b}| = 1$, the usual dot product in $\mathbb{R}^3$ is $\mathbf{a} \circ \mathbf{b} = \cos\theta$, where θ is the angle between the spin directions. Figure 8.3 displays the setup of the Stern-Gerlach apparatuses for measuring the transition probability between these two states and the meaning of the transition probability.

Extend the 3D vectors $\mathbf{a}$ and $\mathbf{b}$ to four-vectors a and b, respectively. Then, from (8.14), it follows that

$$\varrho_1 = e_0 + ia, \quad \varrho_2 = e_0 + ib. \tag{8.16}$$

In Fig. 8.4, we represent the normalized states $\hat{\varrho}_1, \hat{\varrho}_2$, corresponding to the states ϱ_1, ϱ_2, respectively, and their conjugates, by using the decomposition (8.15) and (8.16).

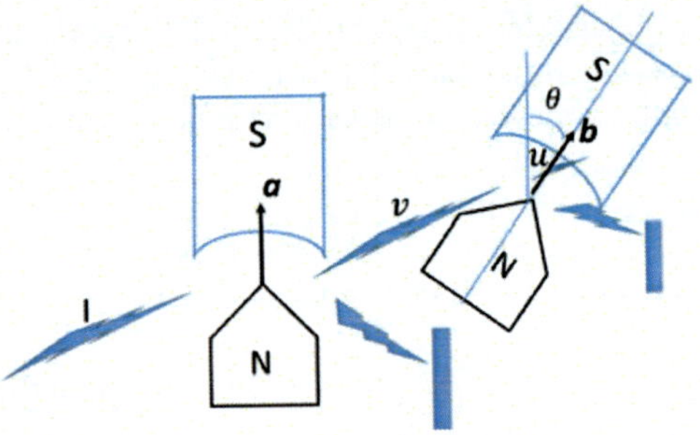

Fig. 8.3 Two Stern-Gerlach apparatuses measure the transition probability between two states. The incoming beam I is in the direction perpendicular to both **a** and **b**. It is split by the first SG apparatus into spin states $|a+\rangle = \varrho_1$ and $|a-\rangle = \bar{\varrho}_1$. The state ϱ_1 is measured by the second SG apparatus, rotated by an angle θ with respect to **a**. We denote by ϱ_2 the state of particles going in the $|b+>$ direction after this measurement. The transition probability $[\varrho_2|\varrho_1]$ is the proportion of particles in state ϱ_1 that passes the second SG apparatus to state ϱ_2

Fig. 8.4 The decomposition (8.15) of normalized spin states $\hat{\varrho}_1 = \frac{1}{\sqrt{2}}(e_0 + ia)$ and $\hat{\varrho}_2 = \frac{1}{\sqrt{2}}(e_0 + ib)$, for states with directions **a** and **b**, respectively, with an angle θ between them

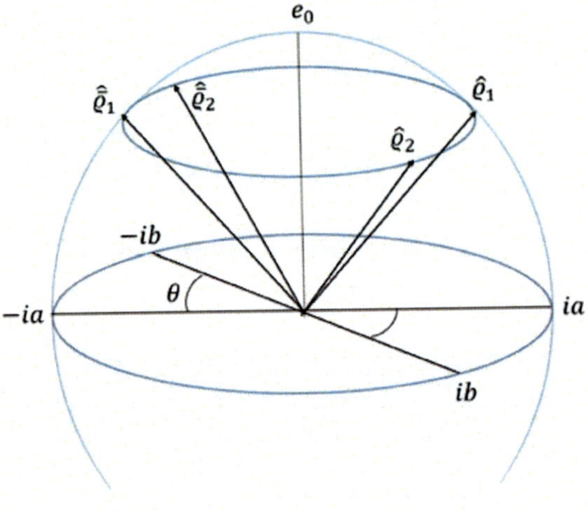

The *transition probability* between two states ϱ_1 and ϱ_2, denoted by $[\varrho_2|\varrho_1]$, is defined in our model by

$$[\varrho_2|\varrho_1] = \hat{\varrho}_2 \star \hat{\varrho}_1. \tag{8.17}$$

This coincides with the usual definition of transition probability in the operator formulation of quantum mechanics. Thus, the transition probability between the states ϱ_1 and ϱ_2 of (8.16) is

$$[\varrho_2|\varrho_1] = \frac{1}{2}\varrho_2 \star \varrho_1 = \frac{1}{2}(1 + \cos\theta) = \cos^2\frac{\theta}{2}, \tag{8.18}$$

which coincides with the known formula for spin transition probability in quantum mechanics. Note that from (8.18), it follows that the value of the transition probability is always between 0 and 1. The transition probability is the proportion of electrons in state ϱ_1 which will pass the filter selecting particles in state ϱ_2.

The spin state $\bar{\varrho}_2$ corresponds to the spin direction direction $-\mathbf{b}$, opposite to the direction of ϱ_2. The meaning of this state is that the rotation of the particle is in the same plane as ϱ_2, but in the opposite direction. Since $\bar{\varrho}_2 = e_0 - ib$, the transition probability between ϱ_1 and $\bar{\varrho}_2$ is

$$[\bar{\varrho}_2|\varrho_1] = \frac{1}{2}\bar{\varrho}_2 \star \varrho_1 = \frac{1}{2}(1 - \cos\theta) = \sin^2\frac{\theta}{2}. \tag{8.19}$$

This shows that the sum of the transition probabilities from ϱ_1 to ϱ_2 and $\bar{\varrho}_2$ is 1, as it should be.

Note that when the directions of rotations of two spin states ϱ_1 and ϱ_2 differ by an angle θ, the transition probability from ϱ_1 to ϱ_2 depends on $\theta/2$. This is due to the fact that the spin state depends also on the four-velocity, and the transition probability is defined only for states with a common four-velocity. See, for example, the states ϱ_1 and $\bar{\varrho}_1$ in Fig. 8.4, representing two electrons rotating in opposite directions about a common axis. The corresponding angular velocities differ by 180°, while the angle between ϱ_1 and $\bar{\varrho}_1$ is 90°. This is due to the common real part e_0.

The vector e_0 represents a state of zero spin and can be decomposed as $e_0 = \frac{1}{2}(\varrho_1 + \bar{\varrho}_1) = \frac{1}{2}(\varrho_2 + \bar{\varrho}_2)$, a sum of two electrons with opposite spin. Thus, this vector may represent the state of a *singlet* [7]—a pair of electrons with opposite spin. Singlets are used to study the properties of the spin of an electron. A singlet can be physically separated into two electrons A and B. Now separately and simultaneously measure the spin of each electron with separate SG apparatuses. If the two apparatuses are in the same direction, then the outcomes will always be opposite. For example, if A is spin up, then B is spin down. This indicates that if the spin state of A after the measurement by the SG apparatus is ϱ_1, then we can know, *even without measuring*, that the state of its companion B is $\bar{\varrho}_1$.

This state of affairs led Einstein to propose that particles have some "hidden variables" which are not measured. Otherwise, how can we instantaneously know the spin of B, even if we are far away from A? Others disagreed with Einstein and claimed that information about the spin in different directions is only probabilistic. Bell devised a criterion to test which approach is correct. He derived an inequality, assuming that the hidden variables approach is valid, for the correlation $C(\theta)$ between the spin state measurements in the directions $\mathbf{a}$ and $\mathbf{b}$ with an angle θ between them. The experiments were performed on two electrons A and B which formed a singlet and were separated without changing their spin. The electron A was measured by an SG apparatus in direction $\mathbf{a}$, and B was measured in the direction $\mathbf{b}$. The correlation $C(\theta)$ is defined to be $P_1 - P_2$, where P_1 is the probability that the two measurements gave the same result, that is, either both were spin up or both were spin down, and P_2 is the probability that the two measurements gave the opposite results, one spin up and the other spin down.

We use the following notation. Let p denote the probability that A is in the state spin up, implying that $1-p$ is the probability that A has spin down. If the measurement of A on apparatus 1 was spin up, we denote its spin state by ϱ_1^A. We infer that if B were to be measured by the same apparatus, its state would be $\bar{\varrho}_1^B$. Similarly, if the measurement of A on apparatus 1 was spin down, we denote its spin state by $\bar{\varrho}_1^A$, and we infer that if B were to be measured by the same apparatus, its state would be ϱ_1^B. The states ϱ_2^B and $\bar{\varrho}_2^B$ are defined similarly for measurements by the second apparatus. Then

$$P_1 = p[\varrho_1^A|\varrho_2^B] + (1-p)[\bar{\varrho}_1^A|\bar{\varrho}_2^B] = p[\bar{\varrho}_1^B|\varrho_2^B] + (1-p)[\varrho_1^B|\bar{\varrho}_2^B] = \sin^2\frac{\theta}{2}.$$

Similarly,

$$P_2 = p[\varrho_1^A|\bar{\varrho}_2^B] + (1-p)[\bar{\varrho}_1^A|\varrho_2^B] = p[\bar{\varrho}_1^B|\bar{\varrho}_2^B] + (1-p)[\varrho_1^B|\varrho_2^B] = \cos^2\frac{\theta}{2}.$$

Thus, the correlation is

$$C(\theta) = \sin^2\frac{\theta}{2} - \cos^2\frac{\theta}{2} = -\cos\theta, \tag{8.20}$$

which agrees with the prediction of quantum mechanics and violates Bell's inequality. Moreover, the value of $C(\theta)$ has been verified experimentally by several experiments. This formula is used in the explanation of experiments testing the validity of quantum mechanics predictions by use of Bell's inequality (see [7, pp. 586, 590]).

The spin state of an electron can be used to describe both its linear and angular momentum. To define angular momentum, we extend the notion of the four-momentum mcw of a non-rotating particle of mass m, moving with four-velocity w, to the four-momentum $p = mc\varrho = mcw + imcRs$ of a moving and rotating particle. With this definition, the real part of the extended four-momentum coincides with the usual four-momentum of a moving particle in special relativity. The imaginary part mcs represents the momentum of angular rotation, called the *spin angular momentum*. In the frame comoving with the particle, the spatial part of the spin angular momentum $\tilde{\mathbf{s}}$, as denoted in (8.1), is

$$\tilde{\mathbf{s}} = mcR\mathbf{s}, \tag{8.21}$$

where $\operatorname{Im}\Lambda^{-1}(\varrho) = (0, R\mathbf{s})$ and Λ is the Lorentz transformation from the lab frame to the comoving frame.

8.6 Motion of Particles with Spin in a Slow-varying Electromagnetic Field

Here, we describe the evolution of a charged spin-$1/2$ particle in a slow-varying electromagnetic field. The field is described, as in Chap. 4, by a linear four-potential $A_\mu(x)$ and the anti-symmetric electromagnetic tensor $F_{\mu\nu}(x)$ derived from it. An electron in a fast-changing field radiates. This radiation is used in synchrotrons, where the electrons with relativistic velocities exposed to a fast-varying electromagnetic field produce strong radiation, called *synchrotron radiation*. The restriction to slow-varying fields is equivalent to ignoring radiation.

A particle under the influence of a slow-varying field changes its state, but remains a single particle and does not generate other particles. The evolution of the four-velocity w of the charge in the field, defined by the anti-symmetric electromagnetic tensor $F_{\mu\nu}(x)$, is described by Eq. (4.27):

$$\frac{dw^\mu(\tau)}{d\tau} = \frac{q}{mc^2} F^\mu_\nu(x) w^\nu(\tau). \tag{8.22}$$

For particles with spin, we extend the previous equation to

$$\frac{d\varrho^\mu(\tau)}{d\tau} = \frac{q}{mc^2} F^\mu_\nu(x) \varrho^\nu(\tau). \tag{8.23}$$

This is the *relativistic equation of motion of a charged particle with spin in an electromagnetic field.* To justify the extended equation, consider a particle with an initial spin state satisfying (8.14). By the claim of section 4.5.1, the particle's spin state satisfies (8.14) for all values of τ. Thus, it is natural to assume that the evolution of the spin state of a charged particle in an electromagnetic field is given by (8.23).

Another indication of the validity of our evolution Eq. (8.23) is the fact that the imaginary part of this equation,

$$\frac{ds^\mu}{d\tau} = \frac{q}{mc^2} F^\mu_\nu s^\nu, \tag{8.24}$$

agrees with the Bargmann, Michel and Telegdi (BMT) equation for the 4D spin vector s (see [61, Sect. 11.11]).

Consider now the motion of an electron initially at rest at $\tau = 0$ in a magnetic field $\mathbf{B}$. Since $d\tau = cdt$, the spatial part of the above equation becomes

$$\frac{d\mathbf{s}}{dt} = \frac{q}{mc} \mathbf{s} \times \mathbf{B}.$$

From Eq. (8.21), the angular momentum of the electron is $\tilde{\mathbf{s}} = mcR\mathbf{s}$. Hence, the previous equation can be rewritten as

$$\frac{d\tilde{\mathbf{s}}}{dt} = qR\mathbf{s} \times \mathbf{B}.$$

From this equation and the definition of the magnetic moment, it follows that the magnetic momentum of an electron is

$$\mu = qR\mathbf{s} = \frac{q}{mc}\tilde{\mathbf{s}}, \tag{8.25}$$

which coincides with (8.1) if the Landé factor $g = 2$. Thus, our model confirms the Uhlenbeck and Goudsmit hypothesis that the Landé factor is $g = 2$.

It is shown in [61, Sects. 11.8 and 11.11] that Eq. (8.24) and $g = 2$ lead to the correct anomalous Zeeman effect and the fine structure splitting.

Chapter 9
The Prepotential

In the previous chapter, we treated the motion, in Minkowski space, of charged particles with spin and derived the transition probabilities between different spin states of spin-$1/2$ particles. This was done by introducing a complex four-velocity to define the state of the particle. The next step is to derive the relativistic evolution equation for such particles in an electromagnetic field. In Chap. 5, the equation of motion (5.4) for charged particles in an electromagnetic field is described by a *real-valued* electromagnetic tensor. Thus, the evolution of the particle's spin state is not included in this equation of motion. Since the Stern-Gerlach experiment demonstrates that a particle's spin does influence its worldline, the treatment of Chap. 5 is not the whole story. To complete the picture of the evolution of test charges, we need to complexify the electromagnetic field.

Electromagnetic fields are generated by moving charges. Most charges also have spin, and their spin has an influence on the field. This can become significant on the microscopic level. From the previous chapter, we have seen that the spin state of a particle uses a complexified spacetime. Thus, there is a need to complexify the electromagnetic field in order to describe the sources.

Another motivation for complexifying the electromagnetic field is simplicity. The standard description of the field by electric and magnetic components has six degrees of freedom. In Chap. 5, the electromagnetic field is described by the linear four-potential $A_\mu(x)$ (5.16), which satisfies the Lorentz gauge and has three degrees of freedom. Now we ask: is this four-potential the gradient of a scalar-valued complex function, which we call a *prepotential*? The answer is yes. We show here that the linear four-potential $A_\mu(x)$ of a field generated by a moving charge is the real part of the gradient of a complex-valued prepotential after a conjugation. This prepotential description has only one complex, or two real, degrees of freedom. Thus, the prepotential is a simpler model for an electromagnetic field. This description is similar to the wave function used in quantum mechanics for the description of particles. In fact, we conjecture that the wave function of a particle is the prepotential of its field.

Y. Friedman and T. Scarr, *A Novel Approach to Relativistic Dynamics*,
Fundamental Theories of Physics 210,
https://doi.org/10.1007/978-3-031-25214-3_9

Whittaker [97] introduced the idea of representing the electromagnetic field by a pair of scalar functions. Ruse [89] improved this description. In [36], it was shown that Whittaker's pair of functions are the real and imaginary part of a complexed-valued prepotential. Barut [8, 9] also described electromagnetic fields by functions. The connection of the prepotential to the Aharonov-Bohm effect [3] is explored in [37]. The approach here was further developed in [43] and [38].

9.1 The Prepotential and the Four-Potential of a Field Generated by a Single Source

In this section, we construct a complex-valued prepotential for an arbitrary (that is, electromagnetic, gravitational, or otherwise) field generated by a single source. We assume only that the field propagates with the speed of light. We then derive the corresponding four-potential, demonstrating that all of the information about the field is contained in the prepotential. We will study the Lorentz invariance and other symmetries of the prepotential and the four-potential.

Let M denote Minkowski space, that is, the space $\mathbb{R}^4$ of spacetime positions, endowed with the Minkowski inner product (3.43) on the set $\widetilde{M}$ of four-vectors. In the inertial lab frame, we use coordinates $x = x^\mu$, where $\mu = 0, 1, 2, 3$.

Consider the field generated by a single source with worldline $\psi(\tau) = \psi^\mu(\tau)$, parameterized by some parameter τ. To define the prepotential at the spacetime point $x = x^\mu$, let

$$r(x) = x - \psi(\tau(x)) \tag{9.1}$$

be the relative position null vector at the retarded time $\tau(x)$ (see Figure 3.14). For convenience, we choose to represent $r(x)$ in bipolar coordinates (see formulas (5.55) in Sect. 5.7)

$$x^0 = \rho^0 \cosh\theta,\quad x^1 = \rho^1 \cos\varphi,\quad x^2 = \rho^1 \sin\varphi,\quad x^3 = \rho^0 \sinh\theta\,. \tag{9.2}$$

Since r is null, it belongs to a hyperplane $\rho^0 = \rho^1 = \rho$, and we write

$$r = \rho(\cosh\theta, \cos\varphi, \sin\varphi, \sinh\theta). \tag{9.3}$$

Note that if $r = 0$, meaning that the point x is on the worldline of the source, then θ and φ are not defined. The backward light cone LC_x about a point x is

$$LC_x = \{x - \rho(\cosh\theta, \cos\varphi, \sin\varphi, \sinh\theta) :\ \rho > 0, -\infty < \theta < \infty, 0 \le \varphi < 2\pi\}. \tag{9.4}$$

The motivation to use bipolar coordinates is as follows. Even though we do not assume that our field satisfies an inverse-square law, it is nevertheless helpful, in terms of dimensional analysis, to observe that the potential of a field obeying an inverse-square law is $U(r) = \alpha/r$, for some constant α defining the strength of the source. This potential has dimensions $\frac{[\alpha]}{\text{length}}$. If this potential is the derivative of a prepotential, then the prepotential should have dimensions $[\alpha]$. That is, the prepotential should be the product of α and a unit-free scalar describing the relative position null vector r. As we know, angles are unit-free scalars associated to vectors. It is this observation which led us to bipolar coordinates, with its two angles.

Next, for any point x with $x^0 \geq 0$, we define a local orthonormal frame $\{d_{\boldsymbol{\alpha}}\}$:

$$d_{\mathbf{0}} = \frac{\partial x}{\partial \rho^0} = (\cosh\theta, 0, 0, \sinh\theta), \quad d_{\mathbf{1}} = \frac{\partial x}{\partial \rho^1} = (0, \cos\varphi, \sin\varphi, 0)$$
$$d_{\mathbf{2}} = \frac{1}{\rho^1}\frac{\partial x}{\partial \varphi} = (0, -\sin\varphi, \cos\varphi, 0), \quad d_{\mathbf{3}} = \frac{1}{\rho^0}\frac{\partial x}{\partial \theta} = (\sinh\theta, 0, 0\cosh\theta)$$

and the dual basis $\{d^{\boldsymbol{\alpha}}\}$ of co-vectors $d^{\boldsymbol{\alpha}}_{\mu} = \eta_{\mu\nu} d^{\nu}_{\boldsymbol{\alpha}}$. In this basis,

$$r = \rho(d_{\mathbf{0}} + d_{\mathbf{1}}). \tag{9.5}$$

We use **boldface** indices for the BP basis and dual basis to distinguish them from the usual basis. Note that differentiation by θ and φ maps the local frame into itself:

$$d_{\mathbf{0},\theta} = d_{\mathbf{3}},\ d_{\mathbf{3},\theta} = d_{\mathbf{0}},\ d_{\mathbf{1},\varphi} = d_{\mathbf{2}},\ d_{\mathbf{2},\varphi} = -d_{\mathbf{1}}. \tag{9.6}$$

As mentioned above, the prepotential of the field is proportional to the strength α of the source and depends on an angle determined by the relative position of the source. Recall from Sect. 8.3 that complexified Minkowski space is $M_c = M \oplus iM$, endowed with the unconjugated Minkowski inner product (8.8). We propose the following definition.

Definition 9.1 For any $x \in M_c$ not on the worldline of the source, the *prepotential of the field generated by a single source* of constant strength α is defined by

$$\psi(x) = \alpha(\theta - i\varphi), \tag{9.7}$$

a product of α and the complex angle

$$\zeta = \theta - i\varphi \tag{9.8}$$

of the relative position $r(x)$ defined by (9.3).

Note that the field strength α can be positive or negative.

It is just as natural to define $\zeta = \theta + i\varphi$. This would amount to a change of orientation of the spatial axes, and, later, would require using right multiplication in our representation of the Lorentz group instead of left multiplication. We have chosen to use left multiplication since operators usually act from the left. Hence, we have chosen a minus sign in the complex angle. This choice is somewhat arbitrary as both left and right multiplication lead to representations of the Lorentz group. In Sect. 9.3, we will show that the prepotential is invariant under these representations.

Note that any stellar observation can be translated into the complex angle (9.8). Let $\tilde{\varphi}, \tilde{\theta}$ be the angles of spherical coordinates for some observation. We may assume that $\varphi = \tilde{\varphi}$, and using (9.3), we obtain $\theta = \tanh^{-1}(\cos\tilde{\theta})$. Thus, ζ is calculable directly from observation.

Since $0 \le \varphi < 2\pi$, the prepotential $\psi(x)$ defined by (9.7) is not continuous. We may, however, define the following continuous prepotential.

Definition 9.2 For any $x \in M_c$ not on the worldline of the source, the *continuous potential of the field generated by a single source* of constant strength α is defined by

$$\tilde{\psi}(x) = \alpha e^{\zeta}, \tag{9.9}$$

where ζ is defined by (9.8).

We show later that for a single-source field, the four-potential derived from Definition 9.1 is a complex extension of the usual electromagnetic four-potential defined by (5.16), while the continuous four-potential is closer to the quantum mechanics model.

The following properties of the derivatives of the complex angle with respect to r^μ will be needed later. Introduce two null-vectors $n = (1, 0, 0, -1)$ and $m = (0, -1, -i, 0)$. These are two of the four-vectors of the Newman-Penrose tetrad [79]. Then $m \cdot n = 0$, $r \cdot n = \rho e^{\theta}$ and $r \cdot m = \rho e^{i\varphi}$, implying that $\zeta = \ln \frac{r \cdot n}{r \cdot m} = \ln(r \cdot n) - \ln(r \cdot m)$. Thus,

$$\frac{\partial \zeta}{\partial r^\mu} = \frac{n_\mu}{r \cdot n} - \frac{m_\mu}{r \cdot m}. \tag{9.10}$$

This implies that

$$\frac{\partial \zeta}{\partial r^\mu} r^\mu = \frac{r^\mu n_\mu}{r \cdot n} - \frac{r^\mu m_\mu}{r \cdot m} = \frac{r \cdot n}{r \cdot n} - \frac{r \cdot m}{r \cdot m} = 0. \tag{9.11}$$

Hence, the gradient of ζ with respect to r is orthogonal to r. Moreover, using (9.10) and the fact that n, m are null, the d'Alembertian of ζ with respect to r vanishes:

$$\Box_r \zeta = \eta^{\mu\nu} \frac{\partial^2 \zeta}{\partial r^\mu \partial r^\nu} = -\frac{n \cdot n}{(r \cdot n)^2} + \frac{m \cdot m}{(r \cdot m)^2} = 0. \tag{9.12}$$

9.2 Representations of the Lorentz Group on M_c

Since the prepotential is a complex-valued function on M, its gradient is a covector in complexified Minkowski space M_c. In order to obtain a representation of the Lorentz group under which the prepotential is invariant, we introduce the following identification of M_c with 2×2 complex matrices:

$$\Phi(x) = \begin{pmatrix} x^0 + x^3 & x^1 + ix^2 \\ x^1 - ix^2 & x^0 - x^3 \end{pmatrix} \tag{9.13}$$

(see [11, 62] and references therein). The components of this matrix are the coordinates of x with respect to the Newman-Penrose null tetrad. However, identifying the vector as a matrix enables the use of additional mathematical tools such as matrix multiplication and determinants. We observe that matrix multiplication is associative but not commutative. Non-commutativity is one of the major distinctions between the classical and quantum models. Note that $(x)^2 = \det \Phi(x)$.

Using the Pauli matrices

$$\sigma_0 = \begin{pmatrix} 1 & 0 \\ 0 & 1 \end{pmatrix}, \ \sigma_1 = \begin{pmatrix} 0 & 1 \\ 1 & 0 \end{pmatrix}, \ \sigma_2 = \begin{pmatrix} 0 & i \\ -i & 0 \end{pmatrix}, \ \sigma_3 = \begin{pmatrix} 1 & 0 \\ 0 & -1 \end{pmatrix}, \tag{9.14}$$

we can write

$$\Phi(x) = x^\mu \sigma_\mu. \tag{9.15}$$

The Pauli matrices satisfy the Canonical Anti-commutation Relations

$$\{\sigma_j, \sigma_k\} = \frac{1}{2}(\sigma_j \sigma_k + \sigma_k \sigma_j) = \delta_{jk} I \quad , \quad j, k = 1, 2, 3, \tag{9.16}$$

where δ_{jk} is the Kronecker delta symbol (3.54) and I is the identity. The multiplication rules for the Pauli matrices are

$$\sigma_0 \sigma_j = \sigma_j \sigma_0 = \sigma_j, \ \sigma_1 \sigma_2 = -i\sigma_3, \ \sigma_2 \sigma_3 = -i\sigma_1, \ \sigma_3 \sigma_1 = -i\sigma_2. \tag{9.17}$$

The identification of M_c with the matrices $\Phi(x)$ is related to the quaternion approach [84], which has long been used in relativity, geometry and quantum mechanics. The quaternions have the form $q = q^0 + q^1\mathbf{i} + q^2\mathbf{j} + q^3\mathbf{k}, q^\mu \in \mathbb{C}$. The multiplication rules (9.17) of the matrices $\sigma_0, i\sigma_1, i\sigma_2, i\sigma_3$ are the same as those of the quaternions $\{1, \mathbf{i}, \mathbf{j}, \mathbf{k}\}$. Note that quaternion multiplication is associative but not commutative, as is matrix multiplication. This means that all of our results here can be translated into the language of the quaternions.

The regular representation π of the Lorentz group on M is by Lorentz boosts B_j in the direction x^j and rotations R_j about the x^j axis, for $j = 1, 2, 3$. This representation has a natural extension to a representation on M_c.

Under the identification (9.13) of M_c with 2×2 matrices, the representation π may be obtained by multiplication of $\Phi(x)$ from both the left and the right by certain 2×2 matrices. This indicates that the regular representation π can be decomposed into a product of two representations π_L and π_R of the Lorentz group on M_c, where π_L acts by left multiplication on $\Phi(x)$ and π_R acts by right multiplication on $\Phi(x)$.

To define a representation of the Lorentz group, it is enough to define the representation of the generators $\mathfrak{B}_j$ of boosts B_j and the generators $\mathfrak{R}_j$ of rotations R_j.

Definition 9.3 Under the representation π_L, the generators $\mathfrak{B}_j$ of boosts in the direction x^j and the generators $\mathfrak{R}_j$ of rotation about the x^j axis are represented as:

$$\pi_L\mathfrak{B}_j(x) = \Phi^{-1}\left(\frac{1}{2}\sigma_j\Phi(x)\right), \quad \pi_L\mathfrak{R}_j(x) = \Phi^{-1}\left(\frac{i}{2}\sigma_j\Phi(x)\right), \tag{9.18}$$

for any $x \in M_c$, where σ_j are the Pauli matrices (9.14).

To show that π_L is a representation of the Lorentz group, it is enough to check that $\pi_L\mathfrak{B}_j$ and $\pi_L\mathfrak{R}_j$ satisfy the same commutation relations as the corresponding generators of the Lorentz group. This follows directly from (9.16) and (9.17).

Under this representation, the boosts B_j act by left multiplication of $\Phi(x)$ by $\exp(\frac{w}{2}\sigma_j)$, and rotations R_j act by multiplication by $\exp(\frac{i\omega}{2}\sigma_j)$, for parameters w, ω. From (9.16), it follows that

$$\exp\left(\frac{w}{2}\sigma_1\right) = \begin{pmatrix} \cosh\frac{w}{2} & \sinh\frac{w}{2} \\ \sinh\frac{w}{2} & \cosh\frac{w}{2} \end{pmatrix}, \quad \exp\left(\frac{i\omega}{2}\sigma_1\right) = \begin{pmatrix} \cos\frac{\omega}{2} & i\sin\frac{\omega}{2} \\ i\sin\frac{\omega}{2} & \cos\frac{\omega}{2} \end{pmatrix}.$$

This implies that

$$\pi_L B_1 = \Phi^{-1}\left(\exp\left(\frac{w}{2}\sigma_1\right)\Phi(x)\right) = \begin{pmatrix} \cosh\frac{w}{2} & \sinh\frac{w}{2} & 0 & 0 \\ \sinh\frac{w}{2} & \cosh\frac{w}{2} & 0 & 0 \\ 0 & 0 & \cosh\frac{w}{2} & -i\sinh\frac{w}{2} \\ 0 & 0 & i\sinh\frac{w}{2} & \cosh\frac{w}{2} \end{pmatrix} \tag{9.19}$$

and

$$\pi_L R_1 = \Phi^{-1}\left(\exp\left(\frac{iw}{2}\sigma_1\right)\Phi(x)\right) = \begin{pmatrix} \cos\frac{\omega}{2} & i\sin\frac{\omega}{2} & 0 & 0 \\ i\sin\frac{\omega}{2} & \cos\frac{\omega}{2} & 0 & 0 \\ 0 & 0 & \cos\frac{\omega}{2} & \sin\frac{\omega}{2} \\ 0 & 0 & -\sin\frac{\omega}{2} & \cos\frac{\omega}{2} \end{pmatrix}. \tag{9.20}$$

The formulas for the boosts B_2, B_3 and rotations R_2, R_3 are similar. This establishes that π_L is a spin-1/2 representation of the Lorentz group.

Note that the generator $\mathfrak{B}_j$ of a boost in the direction x^j is associated with the acceleration in this direction, while the generator $\mathfrak{R}_j$ of rotation about the x^j axis is associated with the corresponding rotational acceleration. Thus, it is natural for the generator of a rotation to be i times the generator of a boost. Since the representation π_L acts by left multiplication on the matrix $\Phi(x)$, it acts linearly on the columns of $\Phi(x)$. Hence, this representation has two invariant subspaces

$$M_1 = \{x : x^1 + ix^2 = 0, x^0 - x^3 = 0\}, \quad M_2 = \{x : x^0 + x^3 = 0, x^1 - ix^2 = 0\}, \tag{9.21}$$

corresponding to the first and second column, respectively.

The matrix representation of the generators (9.19) of boosts are the Majorana-Oppenheimer matrices (see [23]).

$$(K_1)^\alpha_\beta = \frac{1}{2}\begin{pmatrix} 0 & 1 & 0 & 0 \\ 1 & 0 & 0 & 0 \\ 0 & 0 & 0 & -i \\ 0 & 0 & i & 0 \end{pmatrix}, \quad (K_2)^\alpha_\beta = \frac{1}{2}\begin{pmatrix} 0 & 0 & 1 & 0 \\ 0 & 0 & 0 & i \\ 1 & 0 & 0 & 0 \\ 0 & -i & 0 & 0 \end{pmatrix}, \quad (K_3)^\alpha_\beta = \frac{1}{2}\begin{pmatrix} 0 & 0 & 0 & 1 \\ 0 & 0 & -i & 0 \\ 0 & i & 0 & 0 \\ 1 & 0 & 0 & 0 \end{pmatrix}.$$

The generators of rotations are iK_j, $j = 1, 2, 3$. (After lowering the upper index, these become antisymmetric matrices).

We can define a dual representation π_R by replacing the multiplication in (9.18) from the left by Pauli matrices σ_j with multiplication by their complex conjugates $\bar{\sigma_j}$ from the right, and replacing i with $-i$ for the generators of rotations. This representation can also be defined via the matrices $\overline{K}_j$, which are the complex conjugates of K_j. The matrices $\overline{K}_j$ are antisymmetric on M_c. The invariant subspaces of π_R are the two subspaces corresponding to the rows of $\Phi(x)$. These representations commute, and the usual representation π of the Lorentz group is their product: $\pi = \pi_L \pi_R$.

We now explain the meaning of the parameter w in formula (9.19). A direct calculation shows that $\pi B_1 = \pi_L(B_1)\pi_R(B_1) = \pi_L(B_1)\overline{\pi_L(B_1)}$, which is the matrix of a Lorentz boost corresponding to the velocity $\mathbf{v} = (c \tanh w, 0, 0)$. Thus, w is the rapidity of $\mathbf{v}$, and $w/2$ is the rapidity of the symmetric velocity of v (see [45]). The operator $\pi_L B_1$ contains the usual boost in the (x^0, x^1) plane, but with boost velocity equal to the symmetric velocity. An additional boost of the same magnitude acts on the (x^2, ix^3) plane, which is the orthogonal complement of the (x^0, x^1) plane. In the (x^2, ix^3) plane, the Minkowski metric has the same signature as in the (x^0, x^1) plane.

The meaning of the parameter ω in formula (9.20) is similar. A direct calculation shows that $\pi R_1 = \pi_L(R_1)\pi_R(R_1) = \pi_L(R_1)\overline{\pi_L(R_1)}$, which is the matrix of a rotation by an angle ω. The operator $\pi_L R_1$ contains a rotation in the (x^2, x^3) plane by the half-angle $\omega/2$. An additional rotation by $\omega/2$ acts on the (ix^0, x^1) plane, which is the orthogonal complement of the (x^2, x^3) plane. In the (ix^0, x^1) plane, the Minkowski metric has the same signature as in the (x^2, x^3) plane.

The above considerations show that the representations π_L and π_R have symmetries not present in the usual representation π. For example, under π, a boost acts only on the plane spanned by the time direction and the direction of the boost, but leaves the orthogonal complement fixed.

To describe the additional symmetry of π_L and π_R, we introduce a wedge product of two vectors $a, b \in M_c$, similar to the wedge product (5.20). Let $\{e_\mu\}$ be a basis of M_c. The antisymmetric operator $e_{\mu\nu}$ is defined by

$$e_{\mu\nu} = e_\mu \wedge e_\nu, \tag{9.22}$$

where, for any $x \in M_c$,

$$(e_\mu \wedge e_\nu)x = e_\mu(e_\nu \cdot x) - e_\nu(e_\mu \cdot x). \tag{9.23}$$

This wedge product is then extended linearly to define $a \wedge b$ for any pair of vectors in M_c.

The Hodge dual operator $\star$ is a linear map defined by

$$\star\, e_{\mu\nu} = \frac{1}{2}\epsilon^{\alpha\beta}{}_{\mu\nu} e_{\alpha\beta}, \tag{9.24}$$

where ϵ is the *rank* 4 *Levi-Civita pseudoscalar* (compare (5.7)). The square of this operator is minus the identity: $\star\star = -I$. To turn this operator into a symmetry, we define

$$\Lambda = i\,\star\,. \tag{9.25}$$

This is the *helicity operator* used in [30] . Obviously, $\Lambda^2 = I$, implying that Λ is a symmetry.

Definition 9.4 An antisymmetric tensor $\mathcal{F}$ is *self-dual* if

$$\mathcal{F} = \Lambda \mathcal{F} \tag{9.26}$$

and *anti-self-dual* if

$$\mathcal{F} = -\Lambda \mathcal{F}. \tag{9.27}$$

For example, the Majorana-Oppenheimer matrices K_j are anti-self-dual. Thus, the generators of the representation π_L of Definition 9.3 are anti-self-dual. Similarly, the generators of the representation π_R are self-dual. The representations π_L and π_R correspond to different helicities.

9.3 Lorentz Invariance of the Prepotential and the Conjugation

We now prove the following

Claim The prepotential $\psi(x)$ defined by (9.7) and the continuous prepotential $\tilde{\psi}(x)$ defined by (9.9) are invariant under the representation π_L. The complex conjugates of $\psi(x)$ and $\tilde{\psi}(x)$ are invariant under the representation π_R. In addition, $\psi(x)$, $\tilde{\psi}(x)$ and their complex conjugates are invariant under scaling.

To prove the claim, first use (9.3) to write $\Phi(r)$ as

$$\Phi(r) = \rho \begin{pmatrix} e^{\theta} & e^{i\varphi} \\ e^{-i\varphi} & e^{-\theta} \end{pmatrix}. \tag{9.28}$$

Since the determinant of $\Phi(r)$ is zero, the first row $R1$ is proportional to the second row $R2$, and the first column $C1$ is proportional to the second column $C2$. Explicitly,

$$R1 = e^{\bar{\zeta}} R2, \quad C1 = e^{\zeta} C2, \tag{9.29}$$

where ζ is defined by (9.8) and $\bar{\zeta}$ is its complex conjugate.

Under the representation π_L, the boosts and rotations act by multiplication from the left of $\Phi(x)$ by $\exp(\sigma_j\omega/2)$ and $\exp(i\sigma_j\omega/2)$, respectively, for a parameter ω. Since this operation is a linear map on the columns of $\Phi(x)$, the relation (9.29) between the columns is preserved. This implies that e^{ζ}, and hence ζ, are preserved under the representation π_L. Thus, both the continuous prepotential $\tilde{\psi}$ defined by (9.9) and the prepotential $\psi(x)$ defined by (9.7) are invariant under the representation π_L.

Similarly, the representation π_R, acting by multiplication from the right of $\Phi(x)$, preserves the relation between the rows, implying that $\bar{\zeta}$ is preserved under the representation π_R. Thus, the complex conjugates of the prepotential and the continuous prepotential are invariant under the representation π_R.

Since scaling is equivalent to multiplication of the matrix $\Phi(x)$ by the scale factor, scaling preserves the ratio (9.29) between the rows and the columns. Thus, the prepotential, the continuous prepotential and their conjugates remain the same after scaling. This proves the Claim.

The complex electromagnetic field strength tensor $\mathcal{F}$ should be the derivative of a covector-valued four-potential $\mathcal{A}$, meaning that

$$\mathcal{F}_{\mu\nu} = \mathcal{A}_{\nu,\mu} - \mathcal{A}_{\mu,\nu}.$$

We cannot choose $\mathcal{A} = \nabla\psi$, since the curl of a gradient is zero. Instead, we introduce a linear conjugation $\sharp$ on complexified spacetime M_c and define the four-potential to be $\sharp\nabla\psi$.

The conjugation $\sharp : M_c \to M_c$ is defined to be

$$\sharp x = \Phi^{-1}\left(\Phi(x)\sigma_3\right). \tag{9.30}$$

Clearly, $\sharp^2 = I$, and the matrix $\Phi(\sharp x)$ differs from the matrix $\Phi(x)$ only by a change of sign in the second column.

For the regular representation π of the Lorentz group, there are two invariant subspaces, $\mathrm{Re}\, M_c$ and $\mathrm{Im}\, M_c$. Complex conjugation changes the sign of the second invariant subspace. Similarly, for the representation π_L, there are two invariant subspaces, M_1 and M_2, defined by (9.21), and the conjugation $\sharp x$ changes the sign of the second invariant subspace.

Claim The conjugation $\sharp$ commutes with the representation π_L of the Lorentz group.

Since matrix multiplication is associative, and $\sharp$ acts by multiplication of $\Phi(x)$ from the right, while the representation π_L acts by multiplication of $\Phi(x)$ from the left, the conjugation $\sharp$ commutes with the action of π_L. This proves the Claim.

Similarly, a conjugation of x which changes the sign of the second row of $\Phi(x)$ is invariant under the representation π_R.

Under the Pauli matrix representation, the BP basis is

$$\Phi(d_0) = \begin{pmatrix} e^{\theta} & 0 \\ 0 & e^{-\theta} \end{pmatrix}, \quad \Phi(d_1) = \begin{pmatrix} 0 & e^{i\varphi} \\ e^{-i\varphi} & 0 \end{pmatrix},$$

$$\Phi(d_2) = \begin{pmatrix} 0 & ie^{i\varphi} \\ -ie^{-i\varphi} & 0 \end{pmatrix}, \quad \Phi(d_3) = \begin{pmatrix} e^{\theta} & 0 \\ 0 & -e^{-\theta} \end{pmatrix}.$$

Thus,

$$\sharp d_0 = d_3, \ \sharp d_3 = d_0, \ \sharp d_1 = id_2, \ \sharp d_2 = -id_1. \tag{9.31}$$

On the dual basis of covectors, defined by $d^{\alpha}_{\mu} = \eta_{\mu\nu} d^{\nu}_{\alpha}$, $\alpha = 0, 1, 2, 3$, the conjugation is

$$\sharp d^0 = -d^3, \ \sharp d^3 = -d^0, \ \sharp d^1 = id^2, \ \sharp d^2 = -id^1. \tag{9.32}$$

9.4 The Four-Potential of a Moving Source

We now define the complex four-potential of a moving source.

Definition 9.5 The *complex four-potential* $\mathcal{A}$ is defined by

$$\mathcal{A} = \sharp \nabla \psi, \tag{9.33}$$

where ψ is defined by Definition 9.1 and $\sharp$ is defined by (9.30).

To derive an explicit formula for the complex four-potential of a moving source, we first recall (see formula (3.87)) that the derivative of the relative position $r(x)$ is

$$r^{\nu}_{,\mu} = \delta^{\nu}_{\mu} - \frac{w^{\nu} r_{\mu}}{r \cdot w}. \tag{9.34}$$

Since $(r)^2 = 0$ and $(w)^2 = 1$, the inner product $r \cdot w$ is always nonzero, so Eq. (9.34) is always well defined.

Equations (9.5) and (9.6) yield

$$r_{,\mu} = \rho_{,\mu} d_0 + \rho_{,\mu} d_1 + \rho \varphi_{,\mu} d_2 + \rho \theta_{,\mu} d_3. \tag{9.35}$$

Taking the dot product of this equation with d_0 and using (9.34) and (9.5) yields

$$\rho_{,\mu} = d_0 \cdot \left(x_{,\mu} - \frac{w r_{\mu}}{r \cdot w} \right) = \frac{d^0_{\mu} \rho((d_0 + d_1) \cdot w) - \rho(d^0_{\mu} + d^1_{\mu})(d_0 \cdot w)}{r \cdot w}. \tag{9.36}$$

Using (9.22), this can be rewritten as

$$\frac{(\nabla \rho)_{\mu}}{\rho} = \eta_{\mu\nu} \frac{(d_{01} w)^{\nu}}{r \cdot w}. \tag{9.37}$$

Taking now the dot product of Eq. (9.35) with d_3 and d_2, respectively, and using (9.34) and (9.5), we have

$$\rho\theta_{,\mu} = d_3 \cdot \left(\frac{wr_\mu}{r \cdot w} - x_{,\mu}\right) = \frac{\rho(d_\mu^0 + d_\mu^1)(d_3 \cdot w) - d_\mu^3 \rho((d_0 + d_1) \cdot w)}{r \cdot w},$$

or

$$\rho(\nabla\theta)_\mu = \eta_{\mu\nu}\rho\frac{(((d_0 + d_1) \wedge d_3)w)^\nu}{r \cdot w} = \eta_{\mu\nu}\frac{((r \wedge d_3)w)^\nu}{r \cdot w}, \tag{9.38}$$

and

$$\rho\varphi_{,\mu} = d_2 \cdot \left(\frac{wr_\mu}{r \cdot w} - x_{,\mu}\right) = \frac{\rho(d_\mu^0 + d_\mu^1)(d_2 \cdot w) - d_\mu^2 \rho((d_0 + d_1) \cdot w)}{r \cdot w},$$

or

$$\rho(\nabla\varphi)_\mu = \eta_{\mu\nu}\rho\frac{(((d_0 + d_1) \wedge d_2)w)^\nu}{r \cdot w} = \eta_{\mu\nu}\frac{((r \wedge d_2)w)^\nu}{r \cdot w}. \tag{9.39}$$

Therefore, the gradient of the prepotential defined by (9.7) is

$$\nabla\psi = \alpha(\nabla\theta - i\nabla\varphi) = \frac{\alpha}{r \cdot w}\eta((d_0 + d_1) \wedge (d_3 - id_2))w,$$

where we have suppressed the indices μ, ν.

Using (9.32), the complex four-potential (9.33) is

$$\mathcal{A} = \sharp\nabla\psi = \alpha\eta\frac{(d_0 - d_1)((d_0 + d_1) \cdot w) - ((d_3 - id_2) \cdot w)(d_3 + id_2)}{r \cdot w}. \tag{9.40}$$

The numerator is a sum of three components. The first term is

$$d_0(d_0 \cdot w) - d_1(d_1 \cdot w) - d_2(d_2 \cdot w) - d_3(d_3 \cdot w) = w,$$

since this is the decomposition of w by the basis. The second term is

$$d_0(d_1 \cdot w) - d_1(d_0 \cdot w) = d_{01}w,$$

which we recognize as the numerator of (9.37). The last term, the only imaginary one, is

$$id_3(d_2 \cdot w) - id_2(d_3 \cdot w) = -id_{23}w.$$

Thus,

$$\mathcal{A} = \frac{\alpha}{r \cdot w}(1 - id_{23})w + \alpha\nabla \ln \rho. \tag{9.41}$$

The real part of the first term is the Liénard-Wiechert potential

$$A = \frac{\alpha}{r \cdot w} w \tag{9.42}$$

of the electromagnetic field of a moving source, with $\alpha = Q/4\pi\epsilon_0$ (compare to (5.16)). That is, the real part of $\mathcal{A}$ properly defines the electromagnetic field of a moving source. The imaginary part is needed to make the field strength anti-self-dual, as we will soon show. The last term is a gauge, since it is the gradient of a scalar function.

If we use the continuous prepotential (9.9), then the continuous complex four-potential is

$$\tilde{\mathcal{A}} = \sharp\nabla\tilde{\psi} = \mathcal{A}\tilde{\psi}. \tag{9.43}$$

In the Dirac equation, the four-potential of the external field is an operator acting on the wave function [91]. Thus, the continuous four-potential is closer to the quantum mechanics model.

9.5 The Symmetry of the Complex Four-Potential

From (9.40), we can write the complex four potential $\mathcal{A}$ as

$$\mathcal{A} = \frac{\alpha}{r \cdot w}\eta(w + (d_{01} - id_{23})w) = A + \eta S\eta A, \tag{9.44}$$

where A is the Liénard-Wiechert potential (9.42) and S is the operator

$$S = d_{01} - id_{23}. \tag{9.45}$$

Using the definition of d_{01} and d_{23}, it is easily verified that S is a symmetry.

Claim Using the Pauli matrix representation, the operator $S = d_{01} - id_{23}$ acts on $x \in M_c$ by

$$\Phi(Sx) = -\Phi(x)\begin{pmatrix} 0 & e^{-\varsigma} \\ e^{\varsigma} & 0 \end{pmatrix}, \tag{9.46}$$

for Φ defined by (9.13) and the complex angle ς defined by (9.8). Moreover, S commutes with the representation π_L.

It is straightforward to verify (9.46) by checking the action of S on the BP basis. Note that ς is Lorentz invariant under the representation π_L. Since this representation acts by left multiplication on $\Phi(x)$, while S acts by right multiplication, the operator S and the operators of π_L commute. This proves the Claim.

The eigenvectors of S corresponding to the eigenvalues ± 1 are spacetime points x for which the columns of $\Phi(x)$ satisfy $C1 = \mp e^{\varsigma}C2$, respectively. For example, using (9.29), the relative position r of the source at the retarded time with respect to the observer is mapped by S to $-r$, which is the relative position of the observer with respect to the source at the retarded time. Since $\Phi(\sharp r)$ is obtained from $\Phi(r)$ by changing the sign of the second column, we have $S(\sharp r) = \sharp r$.

The operator $P = \frac{1}{2}(1+S)$ is a projection on M_c which commutes with π_L and satisfies $Pr = 0$, $P(\sharp r) = \sharp r$. Using (9.46), for any $x \in M_c$, the norm of $P(x)$ is

$$\det \Phi(P(x)) = \frac{1}{4} \det \Phi(x) \det \begin{pmatrix} 1 & -e^{-\zeta} \\ -e^{\zeta} & 1 \end{pmatrix} = 0.$$

Note that from the definition of S, it follows that $Sw \cdot w = 0$ for any four-vector w. If w is the four-velocity of the source, then this implies that Sw in purely spatial in the frame comoving to the source. We have thus proven the following claim.

Claim The complex four potential $\mathcal{A}$, defined by (9.44), is a scalar multiple of the null vector $w + Sw$. In the frame comoving to the source, the norms of the time and space components of $w + Sw$ are 1 and -1, respectively.

Moreover, $\mathcal{A}$ is twice the projection P of the Liénard-Wiechert potential A of the field of a single source.

9.6 The Prepotential and the Wave Equation

We show now that a single-source prepotential $\psi(x)$ satisfies the wave equation. Using (3.89) and (3.90), it is straightforward to show that

$$\nabla \frac{1}{r \cdot w} = \eta \frac{-w}{(r \cdot w)^2} + \eta \frac{r - (r \cdot a) r}{(r \cdot w)^3}, \tag{9.47}$$

where a is the acceleration of the source at the retarded time.

Claim The single-source prepotential $\psi(x)$, defined by (9.7), satisfies the wave equation $\Box \psi(x) = 0$ for any x outside the source.

Since $\psi(x)$ is proportional to the complex angle $\zeta(x) = \theta - i\varphi$, it is enough to show that $\Box \zeta(x) = 0$. We have

$$\Box \zeta(x) = \eta^{\mu\nu} \frac{\partial}{\partial x^\nu} \left(\frac{\partial \zeta}{\partial r^\kappa} \frac{\partial r^\kappa}{\partial x^\mu} \right) = \eta^{\mu\nu} \frac{\partial^2 \zeta}{\partial r^\kappa \partial r^\alpha} \frac{\partial r^\kappa}{\partial x^\mu} \frac{\partial r^\alpha}{\partial x^\nu} + \eta^{\mu\nu} \frac{\partial \zeta}{\partial r^\kappa} \frac{\partial^2 r^\kappa}{\partial x^\mu \partial x^\nu}. \tag{9.48}$$

From (9.34) and the fact that r is null, we have

$$\eta^{\mu\nu} \frac{\partial r^\kappa}{\partial x^\mu} \frac{\partial r^\alpha}{\partial x^\nu} = \eta^{\kappa\alpha} - \frac{w^\kappa r^\alpha + w^\alpha r^\kappa}{r \cdot w},$$

and from (9.12) and the symmetry of mixed partial differentiation, the first term in (9.48) is

$$\eta^{\mu\nu}\frac{\partial^2\zeta}{\partial r^\kappa\partial r^\alpha}\frac{\partial r^\kappa}{\partial x^\mu}\frac{\partial r^\alpha}{\partial x^\nu}=\eta^{\kappa\alpha}\frac{\partial^2\zeta}{\partial r^\kappa\partial r^\alpha}-2\frac{\partial^2\zeta}{\partial r^\kappa\partial r^\alpha}\frac{w^\kappa r^\alpha}{r\cdot w}=-2\frac{w^\kappa r^\alpha}{r\cdot w}\frac{\partial^2\zeta}{\partial r^\kappa\partial r^\alpha}.$$

Using (9.34) once more, we obtain

$$\eta^{\mu\nu}\frac{\partial^2 r^\kappa}{\partial x^\mu\partial x^\nu}=\eta^{\mu\nu}\frac{\partial}{\partial x^\nu}\frac{\partial}{\partial x^\nu}\frac{\partial r^\kappa}{\partial x^\mu}=\eta^{\mu\nu}\frac{\partial}{\partial x^\nu}\left(\delta^\kappa_\mu-\frac{w^\kappa r_\mu}{r\cdot w}\right)=-\eta^{\mu\nu}\frac{\partial}{\partial x^\nu}\frac{u^\kappa r_\mu}{r\cdot w}.$$

Now use (9.47) and (3.89) to obtain

$$\frac{\partial}{\partial x^\nu}\frac{w^\kappa}{r\cdot w}=\frac{a^\kappa r_\nu}{(r\cdot w)^2}-\frac{w^\kappa w_\nu}{(r\cdot w)^2}+\frac{w^\kappa r_\nu(1-(r\cdot a))}{(r\cdot w)^3}.$$

Since $r\cdot r=0$, we obtain

$$\eta^{\mu\nu}\frac{\partial^2 r^\kappa}{\partial x^\mu\partial x^\nu}=-2\left(\frac{a^\kappa r\cdot r}{(r\cdot w)^2}-\frac{w^\kappa w\cdot r}{(r\cdot w)^2}+\frac{w^\kappa r\cdot r(1-(r\cdot a))}{(r\cdot w)^3}\right)=-2\frac{w^\kappa}{r\cdot w}.$$

Finally, using (9.11), we arrive at

$$\Box\zeta(x)=-2\frac{w^\kappa}{r\cdot w}\left(\frac{\partial^2\zeta}{\partial r^\kappa\partial r^\alpha}r^\alpha+\frac{\partial\zeta}{\partial r^\kappa}\right)=-2\frac{w^\kappa}{r\cdot w}\frac{\partial}{\partial r^\kappa}\left(\frac{\partial\zeta}{\partial r^\alpha}r^\alpha\right)=0$$

This proves the Claim.

9.7 The Electromagnetic Field Tensor of a Moving Source and its Self-Duality

All of the above results are valid for any single-source field propagating with the speed of light. The complex four-potential $\mathcal{A}$, derived from the prepotential $\psi(x)$, is a complex extension of the real-valued Liénard-Wiechert potential (5.16). Based on the ideas of Chap. 8, the imaginary part of $\mathcal{A}$ may be used to model the dynamics ensuing from the spin of the source.

The complex electromagnetic field tensor $\mathcal{F}$ can be defined, as in Chap. 5, from the four-potential $\mathcal{A}$ via

$$\mathcal{F}_{\alpha\beta}=\mathcal{A}_{\beta,\alpha}-\mathcal{A}_{\alpha,\beta}. \tag{9.49}$$

The real part $F_{\alpha\beta}(A)=A_{\beta,\alpha}-A_{\alpha,\beta}$ coincides with the electromagnetic tensor (5.5), which, as shown in section 5.3, properly defines the electromagnetic field of a moving charge. Moreover, the complex extension $\mathcal{F}$ has remarkable symmetry properties.

Claim The tensor $\mathcal{F}(x)$, defined by (9.49), with $\mathcal{A}$ defined by (9.33), is anti-self-dual.

To see this, use (9.31) to write

$$\mathcal{A} = \sharp\nabla\psi = (\psi_{,3}, i\psi_{,2}, -i\psi_{,1}, \psi_{,0}). \tag{9.50}$$

From this, it follows that

$$\mathcal{F}_{10} = \mathcal{A}_{0,1} - \mathcal{A}_{1,0} = \psi_{,31} - i\psi_{,02}, \quad \mathcal{F}_{32} = \mathcal{A}_{2,3} - \mathcal{A}_{3,2} = -i\psi_{,13} - \psi_{,20},$$

which implies $\mathcal{F}_{23} = -i\mathcal{F}_{01}$. Similarly,

$$\mathcal{F}_{20} = \mathcal{A}_{0,2} - \mathcal{A}_{2,0} = \psi_{,32} + i\psi_{,10}, \quad \mathcal{F}_{13} = \mathcal{A}_{3,1} - \mathcal{A}_{1,3} = \psi_{,01} - i\psi_{,23},$$

which implies $\mathcal{F}_{31} = -i\mathcal{F}_{02}$. But

$$\mathcal{F}_{30} = \mathcal{A}_{0,3} - \mathcal{A}_{3,0} = \psi_{,33} - \psi_{,00} \quad \text{and} \quad \mathcal{F}_{21} = \mathcal{A}_{1,2} - \mathcal{A}_{2,1} = i\psi_{,22} + i\psi_{,11}.$$

This implies that $\mathcal{F}_{12} = -i\mathcal{F}_{03}$ if and only if

$$\Box\psi(x) = \psi_{,00} - \psi_{,11} - \psi_{,22} - \psi_{,33} = 0,$$

that is, if and only if ψ satisfies the wave equation, which is true by the claim of the previous section. Therefore, by the definition (9.26) of anti-self-dualness, the tensor $\mathcal{F}(x)$ is anti-self-dual. This proves the Claim.

We point out that the complex tensor $\mathcal{F}$, derived from the complex four-potential $\mathcal{A} = A + \eta S\eta A$ can be written as

$$\mathcal{F} = F - i \star F, \tag{9.51}$$

where F is the real-valued electromagnetic field tensor (5.24) derived from the Liénard-Wiechert potential A (9.42).

9.8 The Prepotential of a General Electromagnetic Field

In Sect. 5.7, we defined the four-potential of a general electromagnetic field from the four-potential of a single-source field. Here, we define here the *prepotential* of any electromagnetic field.

An electromagnetic field is the sum or integral of the fields generated by its various sources. Likewise, the prepotential is also an integral of the prepotentials of the sources. Note that while the four-potential of a single-source field depends on the magnitude, position and four-velocity of the source at the retarded time, the prepotential is independent of the four-velocity of the source and depends only on its magnitude and position at the retarded time. Thus, for any point $x \in M$, the prepotential will be defined by the distribution and the position of the source charges on the backward light cone LC_x about this point.

Using bipolar coordinates (5.55), any point in LC_x is of the form $x - r$, with the null vector r defined by (9.3). Thus, any point in LC_x is defined by (ρ, θ, φ), with $0 \leq \rho < \infty$, $-\infty < \theta < \infty, 0 \leq \varphi < 2\pi$. As in Sect. 5.7, the spatial volume element is $dv = \rho^2 \cosh\theta d\rho d\theta d\varphi$.

Denote by $\sigma(x)$ the source charge density at x. The prepotential at x is then

$$\psi(x) = \int_{LC_x} (\theta - i\varphi)\sigma(x - \rho(\cosh\theta, \cos\varphi, \sin\varphi, \sinh\theta))dv =$$

$$= \int_0^{2\pi} \int_{-\infty}^{\infty} \int_0^{\infty} (\theta - i\varphi)\sigma(x - \rho(\cosh\theta, \cos\varphi, \sin\varphi, \sinh\theta))\rho^2 \cosh\theta d\rho d\theta d\varphi. \tag{9.52}$$

This is the formula for the prepotential of a general electromagnetic field defined from the distribution of the sources charge densities at the spacetime points x.

The prepotential can also be defined from Maxwell's equations. To derive these equations, note that our four-potential, as shown in Sect. 5.3, satisfies the Lorentz gauge. Thus, the d'Alembertian of the four-potential of the field is the four-current density $J_\mu(x)$ of its sources. In other words, $\Box A_\mu(x) = J_\mu(x)$. Substituting the expression for the four-potential (obtained from the prepotential) in this formula, we obtain Maxwell's equations for the prepotential

$$\sharp\nabla(\Box\psi(x)) = J(x). \tag{9.53}$$

This is actually a third-order differential equation. Nevertheless, it can be transformed into the pair of equations

$$\Box\psi(x) = \phi(x) \quad , \quad \nabla\phi(x) = \sharp J(x) \tag{9.54}$$

by introducing an appropriate function $\phi(x)$. The first equation is Poisson's equation, which is of second order. The second equation is a first-order equation for the potential.

The connection of the electromagnetic field to the prepotential can be expressed through matrices α_j, for $j = 1, 2, 3$, on M_c, defined by

$$\Phi(\alpha_j(x)) = \sigma_j \Phi(x)\sigma_3. \tag{9.55}$$

These are the α-matrices of Dirac. It can be shown [38] that the Faraday vector $\mathbf{F} = \mathbf{E} + i\mathbf{B}$ describing the electromagnetic field can be calculated from the prepotential via

$$F_j(x) = \alpha_j(x)^{\mu\nu}\psi_{,\mu\nu}. \tag{9.56}$$

References

1. M.A. Abramowicz, B. Czerny, J.P. Lasota, E. Szuszkiewicz, Slim accretion disks. Astrophys. J. **332**, 646–658 (1988)
2. R. Adler, M. Bazin, M. Schiffer, *Introduction to General Relativity* (McGraw-Hill, New York, 1975)
3. Y. Aharonov, D. Bohm, Significance of electromagnetic potentials in the quantum theory. Phys. Rev. **115**, 485–491 (1959)
4. V.I. Arnold, *Mathematical Methods of Classical Mechanics* Graduate Texts in Mathematics. (Springer, New York, 1984)
5. Babylonian Talmud, Tractate Kedushin, page 70a
6. V. Baccetti, K. Tate, M. Visser, Inertial frames without the relativity principle. J. High Ener. Phys. **119**, 1–43 (2012)
7. L.E. Ballentine, *Quantum Mechanics. A Modern Development*, 2nd edn. (World Scientific, 2015)
8. A.O. Barut, S. Malin, M. Semon, Electrodynamics in terms of functions over the group $SU(2)$: II. Quantization. Found. Phys. **12**, 521–530 (1982)
9. A.O. Barut, S. Malin, M. Semon, Solution of the basic problems of electrodynamics in the group-space formulation. Il Nuovo Cimento **89**, 64–88 (1985)
10. W.E. Baylis, *Electrodynamics, A Modern Geometric Approach*, Progress in Physics, vol. 17. (Birkhäuser, Boston, 1999)
11. W.E. Baylis, R. Cabrera, J.D. Keselica, Quantum/classical interface: classical geometric origin of fermion spin. Adv. Appl. Clifford Algebras **20**, 517 (2010)
12. K. Bliokh, A. Bekshaev, F. Nori, Dual electromagnetism: helicity, spin, momentum, and angular momentum. New J. Phys. **15**, 033026 (2013)
13. H. Bondi, *Relativity and Common Sense: A New Approach to Einstein* (Dover, New York, 1964)
14. L. Brillouin, *Relativity Reexamined* (Academic, New York, 1970)
15. A.H. Compton, A quantum theory of the scattering of x-rays by light elements. Phys. Rev. **21**, 483–502 (1923)
16. S. Chakrabarti, L.G. Titarchuk, Spectral properties of accretion disks around galactic and extragalactic black holes. Astrophys. J. **455**, 623 (1995)

Y. Friedman and T. Scarr, *A Novel Approach to Relativistic Dynamics*,
Fundamental Theories of Physics 210,
https://doi.org/10.1007/978-3-031-25214-3

17. C. Cohen-Tannoudji, B. Diu, F. Laloë, *Quantum Mechanics*, vol. 2 (Wiley, 2006). (ISBN 978-0-471-56952-7)
18. E.U. Condon, H. Odishaw, *Handbook of Physics* (McGraw-Hill, New York, 1958)
19. T. Damour, N. Deruelle, General Relativistic Celestial Mechanics of binary systems I. The post-Newtonian motion. Ann Inst Henri Poincare A **43**, 107-132 (1985)
20. P.A.M. Dirac, *(1982) General Theory of Relativity* (Wiley, New York, 1975)
21. P.A.M. Dirac, *Principles of Quantum Mechanics*, International Series of Monographs on Physics 4th edn. (Oxford University Press, 1982), p. 255
22. C. Duarte, The classical geometrization of the electromagnetism. Int. J. Geom. Methods Mod. Phys. **12**, 1560022 (2015)
23. V.V. Dvoeglazov, Apeiron **5**, 69–88 (1998)
24. A.S. Eddington, A comparison of Whitehead's and Einstein's formulas. Nature **113**, 192 (1924)
25. A. Einstein, Zur Elektrodynamik bewegter Körper. Ann. Phys. **17**, 891 (1905)
26. A. Einstein, Die Feldgleichungen der Gravitation, Sitzungsberichte der Preussischen Akademie der Wissenschaften zu Berlin: 844-847 (1915)
27. A. Einstein, Jahrbuch der Radioaktivitaet und Elektronik 4 411 (1907)
28. A. Einstein, *Ideas and Opinions*. ed. by C. Seeling. Translations and revisions by S. Bargmann. (Crown Pub., New York, 1954)
29. A. Einstein, *The Meaning of Relativity* (Princeton University Press, Princeton, New Jersey, 1955)
30. I. Fernandez-Corbaton, X. Zambrana-Pualto, N. Tischler, X. Vidal, M.L. Juan, Molina-Teriza Phys. Rev. Lett **111** 060401 (2013)
31. D. Finkelstein, Past-future asymmetry of the gravitational field of a point particle. Phys. Rev. **110**, 965–967 (1958)
32. H. Fizeau, Sur les hypothèses relatives à l'éther lumineux. Comptes Rendus **33** 349–355
33. Y. Friedman, *Physical Applications of Homogeneous Balls*. Progress in Mathematical Physics, vol. 40 (Birkhauser, Boston, 2004)
34. Y. Friedman, M. Semon, Relativistic acceleration of charged particles in uniform and mutually perpendicular electric and magnetic fields as viewed in the laboratory frame. Phys. Rev. E **72**, 026603-1–10 (2005)
35. Y. Friedman, M. Danziger, The complex Faraday tensor for relativistic evolution of a charged particle in a constant field. PIERS Proc. **4**, 529–533 (2008)
36. Y. Friedman, S. Gwertzman, The scalar complex potential of the electromagnetic field (2009). arXiv:0906.0930
37. Y. Friedman, V. Ostapenko, The complex pre-potential and the Aharonov-Bohm effect. J. Phys. A: Math. Theor. **43**, 405305 (2010)
38. Y. Friedman, The wave-function description of the electromagnetic field. extit J. Phys.: Conf. Ser. (IARD 2012) **437**, 012018 (2013)
39. Y. Friedman, Relativistic Newtonian dynamics under a central force. Europhys. Lett. **116**, 19001 (2016). https://doi.org/10.1209/0295-5075/116/19001
40. Y. Friedman, J.M. Steiner, Predicting Mercury's precession using simple relativistic Newtonian dynamics. Europhys. Lett. **113**, 39001 (2016). https://doi.org/10.1209/0295-5075/113/39001
41. Y. Friedman, S. Livshitz, J.M. Steiner, Predicting the relativistic periastron advance of a binary without curving spacetime. Europhys. Lett. **116**, 59001–59006 (2016)
42. Y. Friedman, Unification of the Laws of Nature by Refining Newtonian Dynamics, in *12th Miami International Conference on Torah and Science, Florida 2017*, *B'ohr Ha'torah*, vol. 26, pp. 36–45. www.https://www.jct.ac.il/en/publications/bor-hatorah/bor-hatorah-26
43. Y. Friedman, D.H. Gootvilig, T. Scarr, The pre-potential of a field propagating with the speed of light and its dual symmetry. Symmetry **11**(12), 1430 (2019). https://doi.org/10.3390/sym11121430
44. Y. Friedman, T. Scarr, J.M. Steiner, A geometric relativistic dynamics under any conservative force. Int. J. Geom. Methods Mod. Phys. **16**, 1950015 (2019)

45. Y. Friedman, T. Scarr, Symmetry and special relativity. Symmetry **11**, 1235–1249 (2019)
46. Y. Friedman, A physically meaningful relativistic description of the spin state of an electron. Symmetry **13**(10), 1853 (2021). https://doi.org/10.3390/sym13101853
47. Y. Friedman, A.M. Peralta, Representation of symmetry transformations on the sets of tripotents of spin and Cartan factors. Anal. Math. Phys. **12**, 37 (2022). https://link.springer.com/article/10.1007%2Fs13324-021-00644-8
48. Y. Friedman, A unifying physically meaningful relativistic action. Sci. Rep. **12**, 10843 (2022). https://www.nature.com/articles/s41598-022-14740-7
49. G. Galilei, Dialogue Concerning the Two Chief World Systems (1632)
50. W. Gerlach, O. Stern, Der experimentelle Nachweis der Richtungsquantelung im Magnetfeld. Zeitschrift für Physik. **9**(1), 349–352 (1922)
51. S. Ghosh, T. Sarkar, A. Bhadra, Newtonian analogue of corresponding space-time dynamics of rotating black holes: implication for black hole accretion. Mon. Not. R. Astron. Soc. **445**, 4463–4479 (2014)
52. S. Ghosh, T. Sarkar, A. Bhadra, Exact relativistic Newtonian representation of gravitational static spacetime geometries. Astrophys. J. **828**, 6–9 (2016)
53. G. Gibbons, C.M. Will, On the multiple deaths of Whitehead's theory of gravity Stud. Hist. Philos. Mod. Phys. **39**, 41–61 (2008)
54. H. Goldstein, C. Poole, J. Dafko, *Classical Mechanics*, 3rd edn. (Addison-Wesley, 2000)
55. D. Griffiths, *Introduction to Electrodynamics*, 3rd edn. (Dorling Kindersley, New Delhi, India, 1999)
56. Ø. Grøn, H. Sigbjørn, *Einstein's General Theory of Relativity: With Modern Applications in Cosmology* (Springer, 2007)
57. Ø. Grøn, *Lecture Notes on the General Theory of Relativity* (Springer, New York, 2009)
58. J.F. Hawley, S.A. Balbus, The dynamical structure of nonradiative black hole accretion flows. Astrophys. J. **573**, 738–748 (2002)
59. M.P. Hobson, G. Efstathiou, A.N. Lasenby, *General Relativity, An Introduction for Physicists* (Cambridge University Press, 2007)
60. L.P. Horwitz, R. Arshansky, On relativistic quantum theory for particles with spin 1/2. J. Phys. A: Math. Gen. **15**, L659 (1982)
61. J.D. Jackson, *Classical Electrodynamics*, 3rd edn. (Wiley, 1998). ISBN 978-0-471-30932-1
62. H.P. Jakbsen, M. Vergne, Wave and Dirac and representations of the conformal group. J. Funct. Anal. **24** 52–106 (1977). https://www.overleaf.com/project/61e80acf40396e22062823f5operators,
63. R.P. Kerr, Phys. Rev. Lett. **11** (1963)
64. R.P. Kerr, A. Schild, Some algebraically degenerate solutions of Einstein's gravitational field equations. Proc. Symp. Appl. Math. **17**, 199-209 (American Mathematical Society, Providence, 1965)
65. F. Klein, Riemann und seine Bedeutung für die Entwicklung der modernen Mathematik, In: Klein, F. *Gesammelte mathematische Abhandlungen.* (Springer, Berlin, 1923) (reprinted 1973) 3: 482-497. English translation: Riemann and his significance for the development of modern mathematics. Bull. Am. Math. Soc. **1**, 165–180 (1895)
66. S. Kopeikin, M. Efroimsky, G. Kaplan, *Relativistic Celestial Mechanics of the Solar System* (Wiley-VCH, Berlin, 2011)
67. M. Kramer, I.H. Stairs, R.N. Manchester, M.A. McLaughlin, A.G. Lyne, R.D. Ferdman, M. Burgay, D.R. Lorimer, A. Possenti, N. D'Amico, J.M. Sarkissian, G.B. Hobbs, J.E. Reynolds, P.C.C. Freire, F. Camilo, Science **314**, 97 (2006)
68. L. Landau, E. Lifshitz, *The Classical Theory of Fields* (Addison-Wesley, Reading, Massachusetts, 1959)
69. W.H. Lee, E. Ramírez-Ruiz, Accretion modes in collapsars: prospects for gamma-ray burst production. Astrophys. J. **641**, 961–971 (2006)
70. G. Lochak, *La géométrisation de la physique* (Flammarion, Paris, 1994)
71. D.R. Lorimer, Living Rev. Relat. 11, 20088. http://www.livingreviews.org/lrr-2008-8

72. A.I. MacFadyen, S.E. Woosley, Collapsars: gamma-ray bursts and explosions in "Failed Supernovae." Astrophys. J. **524**, 262–289 (1999)
73. G.B. Malykin, Uspechi FN **170**, 1325–1349 (2000)
74. B. Mashhoon, Gravitoelectromagnetism, in *Reference Frames and Gravitomagnetism* ed. by J.-F. Pascual-Sanchez, L. Floria, A. San Miguel, F. Vicente (World Scientific, 2001) 121-32, gr-qc/0311030
75. R. Matsumoto, S. Kato, J. Fukue, A.T. Okazaki, Viscous transonic flow around the inner edge of geometrically thin accretion disks. Astron. Soc. Jpn. **36**, 71–85 (1984)
76. A. Michelson, E. Morley, On the relative motion of the earth and the luminiferous ether. Am. J. Sci. **34**(203), 333–345 (1887)
77. H. Minkowski, Raum und Zeit (Space and Time). Phys Z. **10**, 75–88 (1908–1909)
78. C. Misner, K. Thorne, J. Wheeler, *Gravitation* (Freeman, San Francisco, 1973)
79. E.T. Newman, R. Penrose, An Approach to gravitational radiation by a method of spin coefficients. J. Math. Phys. **3**, 566–768 (1962)
80. I. Newton, Philosophiæ Naturalis Principia Mathematica (1687)
81. B. Paczyńsky, P.J. Wiita, Thick accretion disks and supercritical luminosities Astron. Astrophys. **88**, 23–31 (1980)
82. A. Papadopoulos, Physics in Riemann's Mathematical Papers. *From Riemann to Differential Geometry and Relativity*, eds. L. ,Ji, A. Papadopoulos, S. Yamada (Springer, Cham, Switzerland, 2017), pp. 151–207
83. W. Pauli, The connection between spin and statistics. Phys. Rev. **58**(8), 716–722 (1940)
84. R. Penrose, W. Rindler, *Spinors and Spacetime*, vol. 1 (Cambridge University Press, 1986)
85. R.V. Pound, G.A. Rebka Jr., Gravitational red-shift in nuclear resonance. Phys. Rev. Lett. **3**(9), 439–441 (1959)
86. K.F. Riley, M.P. Hobson, S.J. Bence, *Mathematical Methods for Physics and Engineering*, 3rd edn. (Cambridge University Press, New York, 2006), p.787
87. W. Rindler, *Relativity, Special, General and Cosmological* (Oxford, New York, 2001)
88. S. Rosswog, E. Ramírez-Ruiz, W.R. Hix, Tidal disruption and ignition of white dwarfs by moderately massive black holes. Astrophys. J. **695**, 404–419 (2009)
89. H.S. Ruse, On Whittaker's electromagnetic "scalar potentials." Q. J. Math. **8**, 148–160 (1937)
90. Stephani et al. *Exact Solutions of Einstein's Field Equations*, 2nd edn. (Cambridge, 2003)
91. A. Sudbery, *Quantum Mechanics and the Particles of Nature* (Cambridge University Press, 1986), p. 318
92. L.H. Thomas, The kinematics of an electron with an axis. Phil. Mag. **7**, 1–23 (1927)
93. G. Uhlenbeck, S. Goudsmit, Spinning electrons and the structure of spectra. Nature **117**, 264–265 (1926)
94. W.A. von Ignatowsky, Einige allgemeine Bemerkungen zum Relativitätsprinzip. Verh. Deutsch. Phys. Ges. **12**, 788–796 (1910)
95. J.M. Weisberg, D.J. Nice, J.H. Taylor, Astrophys. J. 72220101030
96. A.N. Whitehead, *The Principle of Relativity* (Cambridge University Press, 1922)
97. E.T. Whittaker, On an expression of the electromagnetic field due to electrons by means of two scalar potential functions. Proc. Lond. Math. Soc. **2**, 367–372 (1904)
98. C.M. Will, *Theory and Experiment in Gravitational Physics*, revised. (Cambridge University Press, Cambridge, 1993)
99. C.M. Will, The confrontation between general relativity and experiment. Liv. Rev. Rel. **17**, 4 (2014)
100. V. Witzany, C. Lämmerzahl, Pseudo-Newtonian equations for evolution of particles and fluids in stationary space-times. Astrophys. J. **841**, 105–118 (2017)

Index

Y. Friedman and T. Scarr, *A Novel Approach to Relativistic Dynamics*,
Fundamental Theories of Physics 210,
https://doi.org/10.1007/978-3-031-25214-3